A Blueprint for the Hard Problem of Consciousness

Potential States, Strange Loops and The Realization of Information from Cells to Human Consciousness

Authored by

Paulo Jacomo Negro

A Blueprint for the Hard Problem of Consciousness

Authors: Paulo J. Negro

ISBN (Online): 978-1-68108-766-5

ISBN (Print): 978-1-68108-767-2

First published in 2019.

Bentham Science Publishers Ltd.
Executive Suite Y - 2
PO Box 7917, Saif Zone
Sharjah, U.A.E.
Email: subscriptions@benthamscience.net

BENTHAM SCIENCE

CONTENTS

FOREWORD

WHY WE NEED A BLUEPRINT FOR CONSCIOUSNESS

The enigma of how consciousness rises from biological phenomena has perplexed scientists, leading some of the greatest minds, including Nobel Laureates Leon Cooper, Francis Crick, Gerald Edelman, Eric Kandel, and Charles Sherrington, to conclude that answering this question is one of the greatest challenges in science. As Shallice [1] concludes, *"The problem of consciousness occupies an analogous position for cognitive psychology as the problem of language behavior does for behaviorism, namely, an unsolved anomaly within the domain of the approach".*

Similarly, Chalmers [2] states, *"We know consciousness far more intimately than we know the rest of the world, but we understand the rest of the world far better than we understand consciousness".* The puzzle of "consciousness-and-the-brain," or of the "mind-body" problem, is often ranked as one of the top two unanswered scientific questions [3].

When speaking about consciousness, we are referring to its most basic form, the kind falling under the rubrics of 'subjective experience,' 'qualia,' 'sentience,' 'basic awareness,' and 'phenomenal state.' This basic form of consciousness has been best defined by Nagel [4], who claimed that an organism has basic consciousness if there is *something it is like* to be that organism—something it is like, for example, to be human and experience pain, love, or breathlessness. Similarly, Block [5] claimed, *"the phenomenally conscious aspect of a state is what it is like to be in that state".*

The scientific challenge concerning consciousness is far more daunting than what non-experts may surmise: investigators focusing on the problem are not only incapable of having an inkling regarding how something like consciousness could arise from something like the brain, they cannot even begin to fathom how something like consciousness could emerge from any set of real or even hypothetical circumstances. That is, if a neuroscientist were provided with all the materials and dimensions (all eleven of them) of the known universe, he or she would still be unable to have an inkling regarding how to go about having basic consciousness arise from anything physical.

In everyday life, being conscious usually just means "being alive." From the standpoint of this lay intuition, consciousness comes "for free" if one happens to be a living organism: one is alive and, because of this, one is aware of one's surroundings, bodily states, thoughts, and so on. In truth, however, consciousness is actually an achievement of nervous function. To date, we do not have a single clue regarding how consciousness is achieved.

How can something unconscious be turned into something conscious, that is, to something for which there is *something it is like* to be that thing? More specifically, what must we do to a set of unconscious neurons to turn them into something that generates a basic conscious state? We will here pose a straightforward question that I think reveals the gravity of the problem: What would one have to do to cause an (unconscious) machine, such as a robot, to experience, say, a dream? This would require more than just the capacity to detect and respond to external stimuli. The challenge is so daunting that any small glimpse of a new clue regarding how it could be solved is priceless.

In this stimulating treatise, Paulo Negro provides thoughtful insights regarding clues and potential answers to the problem of consciousness-and-the-brain. Through the process, he provides an alternative to currently prevalent approaches, such as Integrated Information Theory, Global Neuronal Workspace Theory, Adaptive Resonance Theory, Orchestrated Objective Reduction, and panpsychism (in which consciousness is a property of all matter), each of which is reviewed thoughtfully in the treatise.

Negro's theory emphasizes action and "the subject and its agency," which, according to Negro predicates any form of conscious experience (*e.g.*, conscious experience of qualities). Negro states, *"There can be no consciousness without a subject because consciousness is the phenomenology of a world presented to a subject."* Unlike prevalent approaches to the study of consciousness, which focus on perception, Negro's account focuses on action: The *"generative model of consciousness has a basis on actions"*. Negro concludes, *"Consciousness provides a device to dynamically process and extract evolutionarily useful information from reality, for stepwise adaptive behaviors. Consciousness is how evolution feels"*.

As noted by Paulo Negro throughout the book, when one embarks on a journey to study consciousness scientifically, one encounters many challenges. Below we delineate just some of these challenges, problems which are addressed throughout this tome.

THE PROBLEM OF CONSCIOUS *VERSUS* UNCONSCIOUS PROCESSES

The natural scientist must explain the difference between conscious processes in the brain and unconscious processes in the brain, of which there are many. Using a *descriptive* approach, the scientist attempts to explain the products of evolution as they are and not as we would have designed them or as they should be (which would be a *normative* approach). (Similar to Negro's account, a descriptive approach focuses on phylogeny). As Negro concludes, *"Natural history is not the search for an abstract theory, but the understanding of the history of life in our planet"*

To the detriment of the scientists, this difference cannot be dismissed and must be explained, one way or another. An explanation is required of science even if one proposes that all matter possesses consciousness (a form of *panpsychism*), for one must explain how unconscious processes could exist in a panpsychist reality. An explanation is required even if one proposes that consciousness serves no role in the nervous system (a form of *epiphenomenalism*). Biological phenomena must be explained and incorporated with the theoretical framework for understanding the rest of the natural work, whether they are deemed to be "useful" or not.

Thus, a complete theory of the natural world must explain the difference between conscious and unconscious processes, a difference that exists in every subfield of psychology and of neuroscience. In perception research, there is the distinction between *supraliminal versus subliminal* processes. In the study of attention, the term 'attentional awareness' is often contrasted with unconscious, 'pre-attentive' processing [6]. In memory research, there is the classic distinction between 'declarative' (explicit) processes and 'procedural' (implicit) processes [7 - 8]. In research on language production and motor control, the conscious aspects of voluntary action and action monitoring are contrasted with the unconscious aspects of motor programming [9 - 10], including the implicit learning of motor sequences [11]. Last, various fields contrast 'controlled' processing, which tends to be associated with consciousness, and 'automatic' processing, which tends to be associated with unconscious mechanisms [12]. In short, the difference between conscious and unconscious nervous processes is inevitable and must be explained, even if one adopts the stances of panpsychism or epiphenomenalism.

THE PROBLEM OF THE UNANALYZABILITY OF CONSCIOUS CONTENTS

As noted by several theorists, basic conscious contents (*e.g*., the color blue, urges, or the smell of lavender) are "unanalyzable," with their qualities being irreducible. Lashley (1923) [13], regarding traditional versions of *subjectivism*, states:

> *"Quality is something unique, indescribable, except in terms of itself. Red is red, green is green. Neither is, by any stretch of the imagination, a form of ether vibration or chemical change in the brain... when by analysis the simplest qualities are reached, nothing more can be said of them save that they are in different, undefinable degrees diverse. They have no describable characters inherent in themselves; they are not analyzable into anything else. They exist by virtue of their indescribable differences and by virtue of nothing else discoverable by introspection (pp. 252 - 253)".*

Thus, the conscious contents of blue, red, a smell, or the urge to blink are the tokens of a mysterious language understood, not by consciousness itself, as is clear in Lashley's conclusion, and not by the physical world. Instead, they seem to be comprehensible by brain systems whose operations are cognitively impenetrable to us. We experience these conscious contents, and they affect behavior and our decision making, but we simply do not know what they are in the sense that we know what other things are (*e.g*., heat). We also know that they do not exist *as such* in the physical, "mental-less" world.

Similarly, our basic sense of a first-person perspective or of *time* is also a creation, creation with no analogs in the world of physical reality [14]. (Regarding the former, most conscious contents appear *as if* from a first-person perspective [15 - 17], be it during waking or dreaming. It has been proposed [15 - 16,18] that the demands of adaptive action selection, in which action targets can be, say, to the right or left of one, require the creation of this first-person perspective, which is a primitive form of 'self'). Regarding the first-person perspective, Negro states, *"Predictive Coding supports a subject-centric view in which actions are not only the source of consciousness, but also the building elements of the subject itself. The neural correlates of consciousness are what the subject is doing"*.

Conscious states such as nausea, the auditory perception of pitch, the urge to sneeze, or the color white [19] simply do not exist in the physical world, just as the symbols displayed in navigational system of a modern car do not exist as such in the physical world. These navigational systems depict restaurants and gas stations as fictional, simplified icons (*e.g*., large forks and gas pumps, respectively). Though fictional, they influence the driver's decisions and actions. Based on this example, one can appreciate that our experience of time, space, nausea, and qualities such as colors are not givens of physical reality, or of being alive, but achievements of devoted and specialized neural activities. These achievements are constructions that, unlike their counterparts in the world of physics, can be manipulated experimentally. Consistent with this view, regarding time, Negro states, *"Local networks affect qualia of time experienced both in visual and auditory experiments"*.

THE PROBLEM OF COMPLETENESS

To the self, the conscious field seems "all encompassing"—complete even when conscious contents are obviously absent, as in the cases of sensory neglect (with anosognosia) and in the phenomenon of change blindness [20]. This "completeness" property is evident also in the dream world, in which circumstances are often irrational and fragmented, but not detected as so. As Paulo Negro concludes, *"conscious experiences can be seen as running models (expectations) simulating reality"* which, to the subject, is the only reality.

This property of completeness stems in part from the fact that *one cannot be aware of that which one is unaware of* [21]. To illustrate this point, Jaynes [21] uses the example of the blind spot in vision. One is usually unaware of the blind spot, simply because—again—one cannot be aware of that which one is unaware of. More generally, if a conscious content is not in the field, then it cannot influence volitional decision-making and action in any way [18]. For example, if the knowledge representations necessary for, say, 'reality monitoring,' are not in the field (*e.g.*, due to high fever), then nothing else can assume the functional influence of these contents. In short, when the appropriate contents are absent, there is no independent repository of knowledge that can supplant these contents [18]. Thus, in the conscious field, there is often the *absence of information* but seldom *information about absence* [20,22]. Consciousness is always the totality of one's experience at one moment in time.

Thus, when the nervous system is functioning normally, the contents of the conscious field at each moment seem not only complete, in the sense described above, but they also seem unambiguous. Merker [23] concludes that, because of the very nature of the constitution of the mechanisms giving rise to conscious sensory representations, the conscious field is actually incapable of representing stimulus ambiguity (*e.g.*, as in the Necker cube), at least at one moment in time.

THE PROBLEM OF CONTEXTUAL SENSITIVITY

In order for action to be adaptive, it must be context-sensitive and must allow for what the Behaviorists called a *conditional discrimination*, in which the response to one stimulus depends on the nature of other stimuli in the stimulus scene. Hence, the consciously experienced meaning of a conscious content in the field, and how one responds to it, depends not only on the semantic representation associated with that conscious content, but also on the relationship between this semantic representation and the semantic representations of the other conscious contents participating in the field at that moment in time [24]. (The critical relationships among stimuli could also be spatial in nature, as in the case in which the adaptive response to *Stimulus A* depends on the spatial distance between, say, *Stimulus B* and *Stimulus C*, with these two stimuli being just two out of many stimuli composing the conscious field). Just as in color perception the nature of the phenomenal experience of one hue is determined in part by the nature of the other hues composing the stimulus scene, a conscious content is framed by the nature of the other conscious contents composing the stimulus scene at that time.

We should add that our intuitions regarding how nervous systems should solve this problem of contextual sensitivity, in which sensory inputs are connected to motor outputs, are actually *computationally impossible* [25 - 26]. The conscious field seems to be solving this problem of contextual sensitivity in a manner that is counterintuitive and different from the manner in which we humans would attempt to engineer a solution to the problem of contextual sensitivity.

It turns out that our intuitions regarding how a nonconscious creature might solve the problem of contextual sensitivity in the perception-and-action cycle are actually computationally impossible. That is, just as a computer cannot solve the traveling salesperson problem, many of the problems humans solve in the perception-to-action cycle are known, *a priori*, to be unsolvable through formal computational means. These problems cannot be solved through computer logic, regardless of how much computational power a computer has.

Nature is doing things differently in its peculiar and often happenstance ways. Reverse engineering the brain and consciousness might reveal clues, not only about what we are, but regarding how to deal with the limitations of our computer and robotic systems, just as the study of the dragonfly, the most sophisticated flyer within and outside the animal kingdom, has led to improvements in aerospace engineering.

Because of their deterministic design, traditional computers are also incapable of generating a truly random event, and hence cannot produce random numbers. Computerized slot machines, for example, are not truly random in nature. But other physical systems, such as those at the quantal level, are truly random in nature [27]. With this in mind, engineers can have slot-machines procure random numbers by having their computers *refer* to a random event (*e.g.*, a decaying radioactive diode). In this case, a computational goal is achieved by having two different kinds of systems interacting with each other, each system benefiting from the physical properties of the other [28].

Likewise, from a descriptive approach to nervous function, one assumption is that the computational goals that humans confront require, given the hardware at hand, at least two kinds of physical processing—one conscious and one unconscious. Until robotic systems can solve the problem of contextual sensitivity (which seems to be computationally impossible), these systems must interact with humans to function adaptively in the natural world, because humans possess a property (a conscious field) that seems capable of somehow solving this challenge of contextual sensitivity.

Thus, in response to the common criticism "but one can imagine that function without consciousness," a scientist today can argue that, first, evolution might not have carried out that function the same way that humans imagine it should be engineered, and, second, that our human ideas regarding how such functions could be carried out nonconsciously are actually impossible, *a priori*.

THE PROBLEM OF THE SPEED OF PROCESSING

Perhaps conscious experiences (*e.g.,* nausea, blue *versus* red) are easier to explain by appealing to, not our physical world, but the bizarre world of neurons, a world operating at vastly different size and speed scales. Much as a classroom model of the solar system that cannot not instantiate the critical phenomenon (strong gravitational fields) that glues our solar system together, the scale of our everyday human experience is far different from that in which consciousness arises (fast neurons interacting with each other). In other words, there is nothing that we can perceive in everyday life that works as fast, and in such an interactive manner, as the functional units of our brain.

Regarding the speed variable, our conscious experience is far slower than whatever processes that give rise to it. Perhaps some property may be arising from the extremely fast speeds of neurons and their networks that cannot happen at the slower speed of our everyday existence, which is the only environment we evolved to understand. The cause and effect relationships among billiard balls striking each other, which we evolved to be capable of understanding, also occurs at a slower scale than the workings of neurons. Returning to the topic of physics, it is important to appreciate that gravity seemed to be a conceptual mystery (dubbed 'action at a distance') until Einstein proved that our basic assumptions about it were wrong: space is not 'empty' but actually has a fabric that is warped by mass, giving rise to the phenomenon we identify as gravity [27]. Our assumptions about how things work (and cannot work) at the 'human scale' of existence might not be applicable to the smaller, faster scale of the world of neurons.

Regarding the speed of neural activity, it is important to appreciate that the puzzle of the "mind-body" problem is the same whether the tokens (conscious contents) differ from each other qualitatively or quantitatively [29]. This is an important notion because we know that the brain can implement codes that are quantitative in nature. Negro emphasizes, *"The brain has a bias towards a rhythmic mode of operation."* Specifically [30], it has been proposed that consciousness depends on, for example, *"precise synchronization of oscillatory neuronal responses in the high frequency range (beta, gamma)"*. Singer [30] adds, *"brain states compatible with conscious processing should be characterized by a high degree of synchrony"*. Negro emphasizes that sustained and widely distributed voltage deflections, increase of beta synchrony and gamma power correspond to a widely distributed cortical activity associated with conscious experiences.

Moreover, a group of neurons receiving information from another brain region cannot "see," in a sense, the anatomical tracts connecting the two regions. Receiving neurons cannot see whether the afference stems from tracts that are visual or auditory. Hence, the receiving neurons must use some other kind of information (perhaps frequencies) to identify whether the information being received is of one kind or another (*e.g.,* visual *versus* auditory).

THE NEED FOR A BLUEPRINT OF CONSCIOUSNESS

We have delineated only a handful of the many problems encountered when attempting to understand consciousness from a scientific point of view. Paulo Negro's blueprint provides the reader with a wonderful journey in which, regarding these and many other challenges, no stone is left unturned and in which priceless clues and thoughtful insights about the true nature of consciousness are revealed, chapter after chapter. To navigate the mounting scientific findings regarding consciousness, and the growing number of theoretical frameworks that attempt to explain these findings, a blueprint for consciousness is essential, now more than ever.

Ezequiel Morsella[1]
Department of Psychology
San Francisco State University, CA
USA

Department of Neurology
University of California, CA
USA

and

Ngoc-Cam T. Bui
San Francisco State University, CA
USA

NOTE

[1]EZEQUIEL MORSELLA is a theoretician and experimentalist who has devoted his entire career to the investigation of differences in the brain between conscious and unconscious circuits that control human action. He is the lead author of the Oxford Handbook of Human Action (2009, Oxford University Press) and was the editor of the Festschrift in honor of Robert M. Krauss. His research has appeared in leading journals in neuroscience and experimental psychology, including Psychological Review and Behavioral and Brain Sciences. After his undergraduate studies at the University of Miami, he carried out his doctoral and postdoctoral studies at Columbia University and Yale University, respectively. He is an Associate Professor in the Department of Psychology at San Francisco State and an Assistant Adjunct Professor in the Department of Neurology at the University of California, San Francisco.

PREFACE

The origin of subjectivity constitutes one of the most challenging scientific questions in any field of human research. The fundamental mechanism that allows physical events to transcend into subjective experiences continues to elude scientists and philosophers, who have aptly named it the Hard Problem of Consciousness.

This blueprint is an abstract model that I envisioned as an explanation for the origin of subjectivity. In many aspects it is still a crude idea demanding testing and validation. It is best seen as a compass, or a map, indicating a direction towards the solution of the problem.

Here, I propose how subjectivity could have arisen out of nothing or, at least, nothing of its own kind. Human beings developed through the course of a conservative natural history. Thus, a connection must exist between the processes that sustain the life of primitive single cells, and the emergence of human consciousness. Moreover, such an association could only have taken place if these ancient biological processes, at some point, started to convey information. In other words, single-celled organisms and their communication channels with the environment generated information, as defined by Shannon's classical information theory.

Current theories of consciousness suffer from an intelligent design bias. Their proponents often search for answers in patterns of information integration, in quantum models, and in other natural phenomena. The subject, a necessary vessel for the existence of subjectivity, becomes an afterthought, and consciousness penetrates every aspect of the physical world, in a cryptic panpsychism.

In contrast, I took the subject's perspective, and looked for a solution to the Hard Problem in the boundaries that separate living subjects from non-living processes. Under this light, consciousness emerges as a form of realization of information, connecting simple life forms to their environment. Consciousness in single-celled organisms, in neural networks, or in the intact human brain shares the same mystery: the nothingness or emptiness of its existence. How could something be there and not be there at the same time?

The realization of information must occur as a self-referential abstraction, otherwise consciousness would not differ from other physical properties of inanimate objects, such as heat.

I use the word "abstraction" to portray a confluence of concepts. In computer science, an abstraction enables programmers to handle the complexity of systems by subsuming details in lower levels. However, there is no icon in a computer screen depicting qualia. The brain extracts and processes regularities from its environment through neuronal assemblies within progressively higher orders. These assemblies have a relational origin, describe phenomenal objects, and can be detached from their conditions of origin. I call this extraction by inference an "abstraction".

Finally, I consider the realization of self-referent information across complexity levels an abstraction. When high-level realization of information occurs by means of a strange loop, this abstraction of self-referent origin can be made available to the organism as a form of subjectivity. Thus, an abstraction is not physically there, despite having a mathematical description. The physical components and interactions that give rise to abstractions can be measured, including the activity of neuron assemblies assessed by electrophysiological experiments or functional brain neuroimaging. Nevertheless, the realization of information cannot be measured directly. It is a form of nothingness.

Consider the following analogy. Capital emerges out of nothing of its order. The concept is well accepted in economic science. Capital is something abstract, not physically present. It is a consequence of the economic dialectic of companies and the market. It exerts an organizing effect on the economic activities of the same companies. Capital can be represented as a variable in economic theory. Capital has a physical and an informational basis, such as in real estate, machine tools, patents, cash, and the know-how of the working force. Capital can be realized as money. Note that paper money also represents a concept: a measure of value, a promise to pay the bearer. In the context of this book, to realize is to make it available at a different level and organization. I use the word realize to express the additional meaning of "knowing[2]". To realize is the act of knowing. The present work shows what sort of mechanisms could render an abstract emergent order, such as capital, aware of itself.

Here, I propose a new version of Hofstadter's strange loop that allows for the realization of information between actual and potential states. I call such states the "information" and "expectation" of processes that connect an organism and its environment. The *information.expectation* relationship underlies an inferential system. This implicit probabilistic model of the environment is encoded in the structure and processes of single-celled organisms and shares the same inferential informational paradigm of abstract information realized across the brain computational hierarchy as the strange loop:

'information.expectation'information.expectation

I studied how local brain networks relate to experiences of awareness. Specifically, I drew ideas from work on spatial localization done on the hippocampus, but also from studies on the awareness of time, interoception – the awareness of our internal body, and the olfactory system. At this level of organization, the brain seems to follow the rules of something called Predictive Coding. The brain does not build perceptions from the bottom up; it creates hypotheses or beliefs from the top down. In other words, Predictive Coding supports the idea of a subject-centric consciousness, and allows for the following reformulation of the strange loop:

'expectation.information'expectation.information

which in the feedback and feedforward language of Predictive Coding of corticocortical and thalamocortical networks can be rewritten as

'top-down.bottom-up'top-down.bottom-up

or

'prediction.predictive-error'prediction.predictive.error

Since the beginning, I believed in a noncomputational origin for consciousness, and built up my thought experiment around this hypothesis. At its roots brain activity is analogic, not digital. Surprisingly, however, the problem of locally encoded consciousness can be approached from a computational perspective. Computation in both silicon chips and the brain follows a hierarchy that inherently supports complex systems. Thus, consciousness conceptually emerges as the realization of information, as a hierarchical abstraction through a new interpretation of strange loops across abstraction levels. This solution explains how local networks, while devoid of consciousness, encode the information necessary for the emergence of consciousness. Furthermore, the idea of consciousness as a computationally derived abstraction fits with the concept of something immaterial springing out of a physical substrate.

This model predicts that the realization of information as some form of awareness requires sustained neuronal rhythms that depict expectation-based Predictive Coding. The awareness of simple phenomenal experiences becomes possible when these oscillatory rhythms integrate sensory information in a pre-motor framework for sequential global actions. Being the subject is the realization of inference inversely mapped out as hidden causes of globally integrated actions. Hofstadter's strange loop applies here.

This action-oriented "generational" model bears computational and noncomputational aspects. It predicts the generation of hypotheses by the brain across multiple scales of cross-frequency oscillatory coupling. A model based on the generation of hypotheses constitutes a simulation in itself. Life and consciousness are thus connected. Inference in single-celled organisms and the integration of information across the computational hierarchies of the human brain share a single principle: the incorporation of the state and causal architecture of the environment.

Furthermore, this blueprint explains how the abstraction we experience as consciousness can emerge from sustained resonant states described by the Adaptive Resonant Theory. Brain resonances may support conscious states by realizing information across multidimensional mathematical objects. Such objects form and disintegrate as self-referential abstractions generated by iterations of local and global neuronal connectivity.

In contrast, I show how the most popular theory that addresses the Hard Problem, the Integrated Information Theory, is fundamentally flawed as an explanation for the cause of consciousness. It is based on a circular reasoning and, despite its qualities; it represents a cryptic form of intelligent design theory.

The ideas contained here occupied my mind as I went for dinner with a friend, who reminded me of a completely different angle to the same scientific search. I followed a thread that connected structures in single-celled organisms, to functional relationships, to informational relationships, to sensory systems, to globalization of information, to realization of information in abstractions, but he asked, "Have you ever thought that people might consider this discussion relevant to their spiritual pursuits?"

Over the years, when engaged in conversations about life after death or the possibility of a soul, I often heard skeptics challenge the spiritual stronghold by asking how a mere abstraction such as the soul could make a neuron fire. My good friend pointed out that, by inverting the thread I had been following, I could actually answer this question. If the transition from physical to abstract goes both ways, then something that lacks any physical structure, the void can indeed make a neuron fire. I am not a religious man, and by no means would I attempt to stretch the reach of this work to the realm of religion. Nonetheless, the system I propose here allows for the generation of abstractions from the concrete, as it allows for the interference of the abstract on the biology of the human brain. I decided to honor and recognize my friend's observation.

In summary, through the course of this book, I outline a blueprint for the solution of one of the most difficult problems in neuroscience and philosophy, the Hard Problem of Consciousness. What started as a narrow question may appeal to a larger audience. Nevertheless, I left my considerations about abstractions as the enablers of physical changes in neurons to an appropriately brief last section, almost as a post hoc. At that point the reader will most likely have made his or her mind about the whole scheme.

Naturally, in studying such a complex problem, I often found myself outside my area of expertise. The neuroscience summarized in this book heavily relies on the work and opinion of major scientists in each different area, who have the credit for all the precise neurobiological depictions. I am fully responsible for any inaccurate interpretation of their perspectives. As I leaned, directly and indirectly on the work of other researchers, I take credit only for the application of their ideas and concepts to test the fitness of the hypotheses advanced by this book.

This book is divided in five sections. I review the Hard Problem of Consciousness in the first section, which includes the shortcomings of current theories, as well as a discussion of the computational and non-computational aspects of the problem. In particular, I explain why the Integrated Information Theory stems from a logical fallacy.

I detail the blueprint in the second section: a variation of Hofstadter's strange loop that uses potential states to explain how something can be generated out of nothing. This section also shows the reason why, from a computational perspective, "embodied abstractions" provide the best representations of conscious experiences. Finally, still in section two, I explain how life and consciousness share the same principles of statistical physics.

I test these hypotheses in the third and fourth sections of the book. I wrote them with no preconceived ideas, other than an intuition about how "processual expectations" underlie the kind of information eventually manifested as human consciousness. I studied local networks in search of the moment when they would become capable of hosting focal aspects of awareness. I was curious about the "why".

I addressed the baffling thought of focal awareness: the brain is empty of consciousness much like local networks and single-celled organisms. The experience of being there that emerges from human consciousness is an illusion of the void. Nevertheless, I address the tension between focal awareness and the global sharing of information that characterizes human consciousness in section four. This section ends with an explanation of how consciousness may emerge even in the absence of the cerebral cortex.

The last section includes the most important thought experiment of this book. The "problem of records" reduces the Hard Problem to a specific set of questions and answers. The "problem of records" is not algorithmically compressible. Thus, it has no computational solution. It necessarily demands a physical, rather than a metaphysical answer. I believe the "problem of records" may eventually allow for a serendipitous solution to the Hard Problem of consciousness. Such a solution may emerge from the symmetry between equations describing abstract "objects" that emerge from cross coupling of neuronal oscillations and the mathematical description of natural phenomena. This section indicates how to connect computational theories of consciousness with theories based on quantum mechanics.

This book can be read in different orders. Readers with a background in biology may wish to start from the sections focused on neuroscience. However, they may find some of the analysis perplexing. For example, it may seem that I treat focal awareness and global consciousness as something concrete. In the initial sections, readers will notice that I do exactly the opposite. I created a variable to depict the elusive "traces" of consciousness across hierarchical levels. The abstractness of the variable allows for the kind of treatment I later apply in the neuroscience review.

Those without a background in biology may find it easier to follow the order of the chapters and start with the philosophical aspects and the review of theories. The discussion starts at an abstract level but becomes practical and applied throughout the neuroscience review. I believe that is likely easier to start from the chapter entitled Final Conclusions, then go through each chapter summary, and finally move to the beginning. This is optimally followed by a new look at the text with emphasis on footnotes. Many footnotes expand on the text and provide clarifying information.

I find the use of terms in the literature confusing and overlapping. In this book, I mostly use the word "awareness" for specific aspects of consciousness, such as awareness of location. I do not use the word "consciousness" to address mental states such as the loss of consciousness during a coma. Here, the word refers to phenomenal consciousness, such as the consciousness of objects, colors, and sounds in an immediate and non-reflexive manner. Higher, global experiences, including the narrative consciousness and self-awareness of human beings, are thus preferentially addressed by the word "consciousness" instead of "awareness".

I would like to thank Dr. Leonardo Ganem for reviewing the manuscript. I am grateful for his advice on science and language. I would like to thank those friends who endured our long and often repetitive conversations about consciousness – even when they had no knowledge of the subject, their support gave me confidence to continue on this journey. You all know who you are. Finally, I would like to thank the reviewers of this manuscript. I do not know who you are, but I thank you for the helpful feedback. I hope the good qualities of this book outweigh its shortcomings.

The subject is void, and the mind its avatar.

Paulo J. Negro, MD PhD
Assistant Clinical Professor of Psychiatry and Behavioral Sciences at the George Washington School of Medicine and Health Sciences
Assistant Professor in the Department of Psychiatry and Behavioral Sciences at Howard University
Chief Medical Officer
Kolmac Outpatient Recovery Centers

CONSENT FOR PUBLICATION

The author consents to the publication of this work.

CONFLICT OF INTEREST

The author(s) declared no conflict of interest regarding the contents of each of the chapters of this book.

ACKNOWLEDGEMENTS

Ideas have pedigree. I would be remiss to not recognize those whose work greatly influenced this publication. György Buzsáki for all things oscillatory. Douglas Hofstadter for the strange loop. Terrence Deacon for the incompleteness of things, Dana Ballard for the hierarchy of abstractions. And Ezequiel Morsella, not only for the action synthesis, but for his kindness. There are many others that influenced this book, as recognized in the text. I thank you all. I should extend my gratitude to Ginger Campbell whose podcast awakened old thoughts and feelings about the origin of human consciousness. And, finally, to Kōun Yamada Roshi whose Gateless Gate still sits on my nightstand.

NOTE

[2]In Italian: “Io ho capito” (I realized), in Portuguese: “Eu compreendi” (I realized or I understood).

DEDICATION

To Amy Morgan for the loving patience

CHAPTER 1

Introduction

Abstract: Hidden pathways allow the emergence of subjective experiences from matter, for the emergence of something out of nothing. David Chalmers aptly named the attempt to unveil such pathways as the Hard Problem of Consciousness. Qualia are the subjective characteristics of experiences, such as the experience of "greenness" elicited by the green color. The Hard Problem of Consciousness corresponds to the search for the origin of qualia. This book divides the Hard Problem in two complementary components: an operation for the realization of awareness and the actual content of this awareness. Although the brains of different species might meet different conditions for the development of species-specific consciousness, all conscious brains share a mechanism that turns matter into experience. This book emphasizes the epistemological distinction between descriptions of the neurological landscapes of species-specific conscious experiences, and a theory that explains consciousness in all kinds of conscious brains.

Keywords: Content of Awareness, Consciousness Definition, Epistemology of Consciousness, Hard Problem Components, Hard Problem Definition, Mechanism of Transcendence, Mechanism for Realization of Awareness, Nothingness, Natural History and Evolutionary Theory Distinction, Physical and Metaphysical Differences, Qualia Definition, Thought Experiment.

WHAT IS THE HARD PROBLEM OF CONSCIOUSNESS

Human consciousness corresponds to a state of self-awareness characterized by three essential features: subjectivity, unity and intentionality [32]. Consciousness bears an intrinsic connection to brain activity, as the quality of human experience directly depends on brain structure and function. Thoughts, perceptions, emotions and memories derive their characteristic human perspective from brain structures, connections and oscillatory patterns.

The origin of subjectivity represents a challenging scientific puzzle: how can experiences emerge from matter? David Chalmers aptly named this question the Hard Problem of Consciousness, which contrasts with easier objective problems comprising tasks such as processing information, focusing attention, controlling behavior, and reporting mental status. Purely physical explanations may not provide a complete solution to the Hard Problem [33 - 35]. An alternative, and

Paulo J. Negro

perhaps better approach involves distinguishing the aspects of the problem that require a physical explanation from those requiring metaphysical explanations. In this blueprint, metaphysical explanations are ascribed an abstract relational origin.

Philosophers often refer to the subjective characteristics of experiences as quale or qualia, in its plural form. The term denotes, for example, the experience of "greenness" or "redness" elicited by the green or red colors. Qualia portray direct experiences, private and intrinsic to the act of experiencing itself. They do not correspond to true characteristics of the world, but to an awareness of the world [36 - 37]. Put in other words, the Hard Problem encompasses the search for the origin of qualia[3]. As a Thought Experiment[4], I chose to divide the Hard Problem in two components: an operation for the realization of awareness and the content of this awareness[5].

CONSCIOUSNESS AS A FORM OF INFORMATION

Subjective experiences emerge from information ordered in material processes. The central problem addressed here refers to the hidden pathways through which something can emerge out of nothing, or at least of nothing of its own kind.

After a certain complexity threshold, a critical mass of brain organization might trigger a primitive form of consciousness. I make no assumptions about a physical property of the brain from which consciousness directly emanates. At this point, the "magical[6]" forces behind such a mechanism do not matter as much as the fact that subjective experiences become possible. I assume the existence of only one fundamental operation behind the emergence of subjective experiences.

THE DIFFERENCE BETWEEN NATURAL HISTORY AND EVOLUTIONARY THEORY

Understanding consciousness from the human brain perspective involves the description of the brain landscape required for the emergence of consciousness, but not necessarily the ultimate cause of consciousness. Different brains might meet different conditions in different species for their consciousness-specific perspectives to emerge, while all conscious brains share a mechanism to transcend matter into experience.

As a comparison, consider our solar system. Understanding our solar system is not the same as creating a general theory of solar systems but knowing about its history and characteristics. Likewise, consider the difference between natural history and evolutionary theory [38]. Natural history is not the search for an abstract theory, but the understanding of the history of life in our planet[7].

NOTES

[3] For an in-depth review of the phenomenology of consciousness the reader should refer to Thomas Metzinger *magnum opus*: Being No One [37].

[4] A gedankenexperiment (Thought Experiment, in German) considers some hypothesis, theory, or principle for the purpose of thinking through its consequences [286].

[5] Realization of awareness addresses how information is transformed across hierarchical levels of brain computation to make it an experience of knowing something. Its content pertains to what is the manifested subjective experience, specifically. For instance, the matching of encoded maps in hippocampal neuronal assemblies with the incoming sensorial experience can signal awareness of current location. The realization of information into this kind of knowing is dependent on sensorial experience, but it is not the sensorial experience itself.

[6] Magic: the power of apparently influencing the course of events by using mysterious or supernatural forces. If magic is to be conjured for the explanation of the Hard Problem, there must be only one magical act.

[7] For an in-depth discussion about the difference between natural history and evolutionary theory the reader should refer to the small but lucid book from Fodor & Piatelli-Palmarini: What Darwin got wrong [38].

CHAPTER 2

Does It Make Sense to Trace Consciousness to Single-Celled Organisms?

Abstract: Popular theories of consciousness convey the notion that traces of consciousness exist even below the level of single-celled organisms. Alternatively, consciousness may emerge from the informational nature of life. Fundamental processes that, when scaled up, might give origin to the "mind" must transit from biological mechanisms to information. Portraying the subject-world dialectic of single-celled organisms as the first exchange that makes this transition possible does not constitute a hypothesis, but a description. Qualia refer to subjective phenomena, and not to something objectively present or to the human brain processes from which such phenomena arise. Nonetheless, qualia can still provide tailored adaptive information about reality from the perspective of the organism. One can reconcile this notion with the illusory nature of qualia by recognizing a variable "Q". This variable represents an abstraction of the processes that transform evolutionarily relevant information into the subject experience. Q represents adaptive inference processes that incorporate the state and the causal architecture of the environment. These inference processes at some point became human consciousness.

Keywords: Adaptive Information, Active Inference, Consciousness in Single-Celled Organisms, Conservative Natural History, Imaginary Numbers, Informational Nature of Life, Membrane Excitability, Origin of Mind, Original Awareness, Panpsychism, Qualia Fundamental Processes, Single-Celled Organisms, Subject-World Dialectic, Traces of Consciousness, Variable Q.

A number of reputable scholars attempt to trace consciousness back to single-celled life forms[8]. In an open letter, Norman Cook, Antonio Damasio, Gil B. Carvalho, Harry T. Hunt and Oliver Sacks [39] responded to Christopher Koch on the nature of consciousness:

> *"That approach entails the search for the properties underlying consciousness down to the level of the protozoan [a single-celled organism] in order to identify the fundamental cell-level mechanisms that, when scaled up in complex nervous systems, give rise to the properties that are typically referred to as "mind". The unanswered question is: What characteristics of living cells lead ultimately to the various, higher-level psychological phenomena that are apparently unique to certain animal organisms? This*

Paulo J. Negro

> *question concerns essentially biological functions—and is distinct from "information processing" approaches that might be implemented in silicon systems."*

Still according to the correspondents:

> *"...the lowest-level candidate mechanism [for consciousness] is membrane 'excitability: the unusual capability of certain types of living cells to sense and respond to stimuli within several milliseconds. (...) The early evolutionary theorists (George John Romanes, Herbert Spencer Jennings, Alfred Binet, and Charles Darwin himself) were duly impressed by the seemingly goal-directed behavior of even single-cell 'animalcules'—despite the complete absence of nervous systems. Their interest was clearly focused on cellular and subcellular phenomena, but a proper understanding of ion flux in producing membrane excitability lay a century ahead. Recent progress in the genetics and physiology of ion channels now indicates both how and why the disruption and restoration of the electrostatic equilibrium of the cellular cytoplasm is so crucial to the unprecedented sentience of both neurons and animalcules. The qualitative argument for considering cation[9] influx to be the 'spark of sentience', as outlined above, is perhaps already plausible, but the quantitative details require further study.".*

In response, Christopher Koch[10] pointed on the importance of information as the foundation for consciousness:

> *"Membrane excitability will play a role here, but as one among many other factors and considerations. Indeed, focusing on ionic channels to understand consciousness, a system-level property par excellence, is as useful as trying to comprehend the nature of the internet by focusing on how electrons flowing onto the gate of a transistor modulate the electric current flow between the other two terminals of the transistor. Yes, transistors are at the root of digital machines' ability to compute. But the properties of the internet that we make daily use of—its near-universal accessibility and its ability to nearly instantaneously retrieve obscure essays, blogs, books, music files, and videos anywhere on the planet—have much more to do with the way information on the Web is indexed and retrieved..."*

Indeed, subjective experiences can arise from some forms of information contained in physical systems, although the processes underlying this association have yet to be unveiled. Consciousness [37] constitutes an abstraction present at certain levels of brain organization and may well bear an association with information, but does it have any discernible traces at levels below? The current scientific search for specific neural correlates of consciousness[11] implies so.

The organization of processes in single-celled organisms within the evolutionary context bears relevance to the question of the origin of subjectivity. Ultimately, consciousness arises from information processing in the human brain. Humans process information at several levels, from subcellular events underlying synaptic[12] activity, to the assembly of large brain regions in an interconnected global workspace. Human brain structure and the processes that generate human consciousness evolved from single-celled organisms through the course of a conservative natural history. Single-celled organisms contain the core components of the synapse and use them in response to environmental stress [41]. Synaptic complexity underlies behavioral diversity, cognitive capacity, and it constitutes an important aspect of the computational complexity of the brain.

Popular theories of consciousness support the notion of traces of consciousness. They use language consistent with the concept of "consciousness-information", such as phi-consciousness [42] or proto-consciousness [43]. Consciousness-information represents a form of information conveying subjectivity across different levels of the brain hierarchical organization. In fact, the authors of these theories refer to information organized at levels even smaller than a single-celled organism. By doing so, they assume a cryptically panpsychist[13] view of the world and have difficulties in delimiting the minimal unit of consciousness.

Alternatively, consciousness may have emerged as a consequence of the informational nature of life. It is possible to transition from membrane excitability and electrostatic equilibrium to forms of information that can be related to the eventual emergence of human consciousness. The subject-world dialectic of single-celled organisms bridging this transition is not a hypothesis. It is a description.

Koch's choice of words in the final part of his response illustrates the importance of a general concept of consciousness-information in the description of primordial forms of consciousness:

> *"Finally, it is quite a different question whether single cell-organisms, worms or other simple metazoans—vastly simpler than mammals with their large brains—have sentience. I do share with the letter writers a hunch that it may well be that 'it feels like something to be a worm'. However, that is a question that right now can't be answered in any meaningful empirically accessible manner."*

QUALIA AS INFORMATION

Subjectivity emerged at an uncertain point during the Earth's natural history. But it did emerge, and natural selection maintained it in the case of our species [44].

Thus, subjective experiences about the world have evolutionary value [45 - 46]. For example, the tridimensional construction of a subjective world through the senses of sight, hearing and touch renders our species more adapted. A phenomenal[14] self [37] that streamlines conscious responses based on subjective experiences about the world has adaptive value.

Consciousness seems to function as a router that flexibly transfers information to and from specialized processors. It belongs to a hybrid system in which serial steps of conscious decision-making organize massive parallel non-conscious computation[15] [46 - 47]. The actions of an organism in the world, *i.e.*, a subject's ongoing relationship with objects, may govern the character of experience [48]. From this perspective, consciousness provides a device to dynamically process and extract evolutionarily useful information from reality, for stepwise adaptive behaviors. Consciousness is how evolution feels.

The term "qualia" refers to certain subjective phenomena, and not to something objectively present or to human brain processes from which such phenomena arise. Qualia do not relate to the true characteristics of the world, but to the world as subjectively experienced; in this sense they may well represent epiphenomena, or noise. Nonetheless, in spite of being a synthetic product of brain activity, qualia can still provide tailored adaptive information that help humans navigate through reality.

Are qualia only mirages created by the organism? Yes and no. The specific quality of a quale should be considered illusory. However, processing reality into quale-type experiences can provide adaptive information. In other words, quale-like experiences convey adaptively organized information from the perspective of the organism[16].

One may reconcile these views by embracing qualia as illusory, while recognizing a variable "Q" as an informational (subjective) perspective about reality. "Q" represents an abstraction of the processes that transform evolutionarily relevant information into the subjective perspective. Conversely, this variable[17] depicts the act of processing reality in a manner that allows for the eventual emergence of qualia.

One cannot measure "Q" in our physical reality[18]. It represents a concept, not different from other mathematical concepts, such as complex numbers[19]. Variables and coefficients are accepted for their usefulness. We do not have direct knowledge of what "Q" is. Q represents adaptive inference processes that incorporate the state and the causal architecture of the environment. It does not even necessarily depict conscious experiences as it portrays "traces" of consciousness across brain hierarchies. Perhaps better stated, it portrays the inference processes at some point realized as human consciousness. The imaginary number i= √ -1 is an accepted mathematical concept, although there is no square root of negative one. If subjectivity is not physical, how to depict it other than with abstract variables?

NOTES

[8] I do not engage in speculation about the boundaries of life or the problems of simulated life for the following reasons: 1) The use of algorithms for life simulation leans towards intelligent design explanations; 2) Despite computational arguments I submit later, so far there is no computational explanation for the origin of subjectivity [61]; and 3) It is unnecessary for the arguments put forward here. I neither advance intelligent design, nor intend to give agency to evolutionary principles. Intelligent design is a weak explanation for life and consciousness. If God created the universe, she did not need designs.

[9] Cation: a positively charged ion.

[10] Christopher Koch is a proponent of the Integrated Information theory (more about this later). Readers interested in his personal (and personable) account of these ideas should refer to his popular book: "Consciousness – confessions of a romantic reductionist" [287].

[11] The content-specific neural correlates of consciousness (NCC) are the neurons or the neuronal mechanisms that determine a particular phenomenal distinction within an experience. The full NCCs are the neural substrates supporting conscious experiences entirely, irrespective of specific contents [192].

[12] The synapse is a structure that allows a neuron to pass signals onto another neuron.

[13] One of the oldest philosophical theories that considers consciousness, mind or soul, as universal features of all things, while also stating that the basic physical constituents of the universe have mental properties.

[14] Phenomenology: the study of structures of experience, or consciousness. "Phenomena": appearances of things, or things as they appear in our experience, or the ways we experience things. Phenomenology studies conscious experience as experienced from a first person point of view [288].

[15] Conscious operations have access to a limited capacity program. They are late in the processing sequence [289].

[16] For example, that we are capable of seeing subjectively perceived colors is adaptive. Certain light wavelengths appear as specific colors such as blue, green or red according to the differential sensitivity of cells in the retina to different light wavelengths. That blue is subjectively seen as blue instead of red is illusory. Note that I am not bringing up the philosophical question of "if blue looks the same to different people", but that for each of us blue looks like blue and not like red. This subject raises the important question of records, which will be elaborated at the end of this book in its own chapter. I believe that the Hard Problem may be solved, perhaps serendipitously, by answering the question of records.

[17] *"In mathematical logic, a variable is either a symbol representing an unspecified term of the theory, or a basic object of the theory, which is manipulated without referring to its possible intuitive interpretation"* [290]. *"Variable: a quantity belonging to a particular mathematical or logical set that can have a number of different values"* [291].

[18] If Q is not a metaphysical concept, but a variable depicting a physical process, the formal identification of its nature will require a change of the physical models of reality.
[19] A complex number is a number that include an imaginary unity that satisfies the equation $i^2 = -1$. Complex numbers extend the concept of the one-dimensional number line to the two dimensional complex plane.

CHAPTER 3

Consciousness: Computable, Noncomputable or Both?

Abstract: The computability issue provides a pivot for understanding the origin of consciousness. A noncomputable origin of consciousness implies that analogic processing of information supported by real numbers and relying on the laws of physics gives rise to subjectivity. In contrast, a computable origin would mean that processing of information relies on algorithms such as those governing digital computing. Neither binary logic nor formal mathematical systems govern the evolution of biological systems. Kurt Gödel formally demonstrated that axiomatic systems do not prove the truth and consistency of all theorems, thus providing evidence that formal mathematical systems cannot fully capture the creative processes that take place in the human brain. However, computational approaches to brain function yield extraordinarily successful results. The operations that underlie the realization of awareness should capture algorithmically compressible regularities. Crucially, silicon and brain computation share a fundamental characteristic: large complex systems must rely on a hierarchical organization where each level becomes increasingly more abstract. Nonetheless, as of yet, no theory has explained how qualia arise from neuro computation. Altogether, these considerations support dividing the Hard Problem into symbiotic computational and non-computational aspects.

Keywords: Abstraction in Computation Science, Algorithms, Axiomatic Method, Brain Operations, Computational Consciousness, EEG Spatiotemporal Patterns, Functional States, Gödel's Proof, Noncomputational Consciousness, Non Computational Subject, Phenomenal and Operational Isomorphism Real Numbers, Phenomenal Transparency, Physical Instantiation, Primitives, Table Lookup, Tractability, Turing Machine.

Researchers often disagree on whether neurocomputation can synthesize the fundamental operations causing subjectivity. Here, I review arguments in defense of both sides of this discussion.

IN FAVOR OF NON COMPUTABILITY

At age 25, Kurt Gödel published a milestone paper in the history of logic and mathematics [49]. Gödel addressed the inherent limitations of the axiomatic

Paulo J. Negro

method, which forms the basis of elementary geometry, a deductive discipline. Geometry relies on the acceptance of certain axioms or propositions, with no need for proof. Other propositions emerge from this system as theorems. Gödel demonstrated that the axiomatic system cannot prove the truth and consistency of all theorems[20] [50].

Gödel's formal demonstration of these limitations provided evidence that formal mathematical systems cannot fully capture the creative processes taking place in the human brain while it solves mathematical problems [50]. Roger Penrose's [51] quantum mechanics account of consciousness sprung from this conclusion. Furthermore, Hofstadter's self-referential "strange loop" theory of consciousness was inspired by Gödel's proof. A version of his "strange loop" will play an important role in the computational explanation of consciousness, later in this book. Nevertheless, whether Gödel's Theorems indicate that the human mind has a non-algorithmic nature, remains a controversial matter [52].

So far, no theory explains how qualia arise from neurocomputation [40]. As classical physics provide no theoretical space for information to become conscious[21], Penrose's turn to quantum mechanics is quite understandable[22]: quantum theory is thought as ultimately explaining everything[23]. The problem is that qualia may not exist as a thing.

Neither binary logic nor formal mathematical systems govern the evolution of biological systems [40,50]. In fact, an abundance of noncomputable mathematical representations of natural phenomena exists [53]. Cicurel and Nicolelis raised the issue of non-tractability. Tractability in digital computation, relates to the number of cycles required to conclude calculations, and to other physical limitations such as energy and memory. These authors emphasized the non-tractability of information processing by the brain. Furthermore, based on the Turing[24] test, they automatically disqualified any proposed digital solution with a processing speed below that of the brain.

IN FAVOR OF COMPUTABILITY

Cicurel and Nicolelis do not make a strong argument for tractability [53]. In reality, the slow firing rate of neurons limits the speed of brain computation. Silicon computers work at clock speeds of 1GHz[25], or a billion operations per second. Neurons operate at a pace more than a million times slower [40].

The brain uses different tricks to make up for the low speed and energy-intensive neuronal firing: massive parallel computation, probabilistic reasoning, encoding information at multiple frequency oscillations, use of special-purpose sensors and effectors and specialized circuits.

The brain computational coding is refined over years of interactions between the body and its environment. Over time, human cortices encode an astonishing amount of momentary experiences that become available on demand, as a table lookup[26]. As an analogy, a Google search produces almost immediate results because the search engine algorithm has previously indexed billions of web pages [40]. Cicurel and Nicolelis compared brain computation with silicon computation [53]. Instead, computation must be understood in the abstract.

A Turing Machine provides a universal model for all computation[27] [40]. It is conceptually independent of a physical substrate. It works by being in a "state", reading a symbol, and following an instruction specifying a new state, an output symbol, and a movement towards a new input symbol (Fig. **1**). The fundamental question is: *"can computation be the root of a theory of brain function?"* [40]

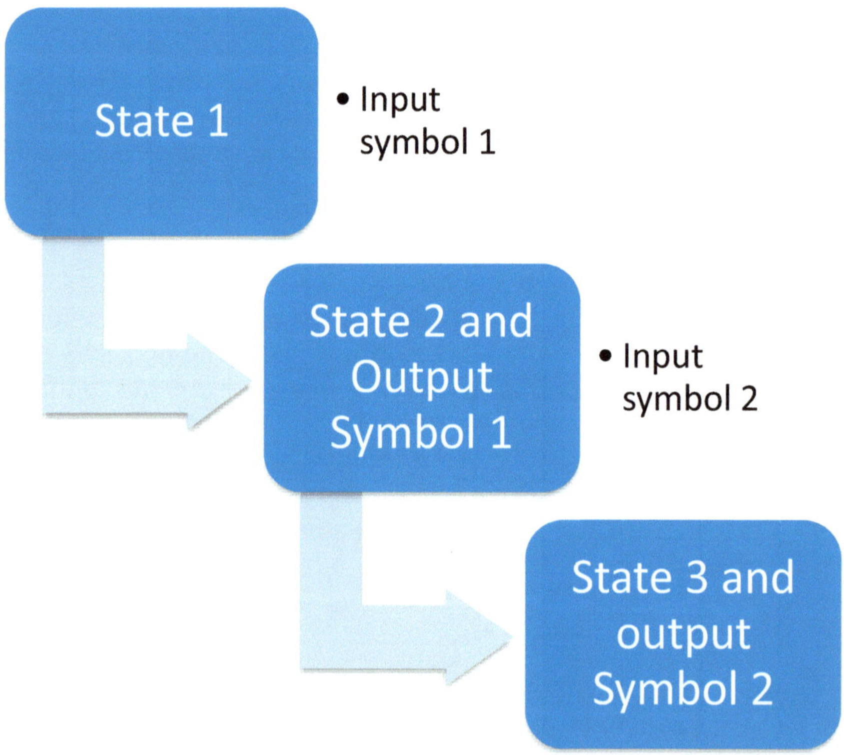

Fig. (1). Abstract representation of a Turing Machine: information is received and processed according to rules of the processor. New information is generated and the processor changes in preparation for receiving more information. What defines computation is not the physical medium, but its abstract architecture.

Computational approaches to brain function yield extraordinarily successful results. Brain programs can be formally represented as states and actions, broken down into steps of approximately 300ms. Tables can provide frameworks that represent different collections of state-action pairs for any behavior encoded in the cortices. Crucially, silicon and brain computation share a fundamental characteristic. As reviewed by Ballard [40], the design of large complex systems must rely on a hierarchical organization where each level becomes successfully more abstract.

Subjectively experienced representations are "transparent" [34]. In other words, a subject will experience the represented object as the object itself without traces of the nonconscious mechanisms that generate the phenomenal experience. Fingelkurts and colleagues [54] proposed that consciousness emerges from a hierarchy of operations[28] among transient neuronal assemblies ranging from "blind" neuronal operations, to cognitive operations supported by elemental functional assemblies (which can be conscious or nonconscious), to complex macro-level operations that support integrated phenomenal experiences.

The electroencephalogram (EEG) has structural and dynamic properties that register and integrate stimuli in a unified and coherent spatiotemporal pattern[29]. Fingelkurts and colleagues developed a methodology to analyze microstates of EEG quasi-stability and the sudden transitions between them[30].

Operations subsuming simpler operations[31] fit a model of abstract descriptions that omit details from lower levels. Thus, as Ballard pointed out, the stability and order provided by increased levels of abstraction may allow brain activity to support formal computation.

Software engineers have historically developed programs from the bottom up, from an original assembly language to more abstract user program languages. Languages with a higher-level of abstraction do not account for all the lower-level representations. The lower levels have details that higher levels suppress. The brain follows the same arrangement: more abstract neural networks lean on less abstract networks [40].

From the perspective of complex systems, brain computation must have hierarchical levels, each with a different set of primitives[32] from the level below. At some point, consciousness conceivably binds lower levels and makes it possible integrated models capable of flexible simulation of self and objects, the past and the future. Primitives are then treated as symbols[33] that can be manipulated [40].

A HYBRID SYSTEM: EMBODIED ABSTRACTIONS

As a thought experiment, I divided the Hard Problem in two components: an operation for the realization of awareness and its ultimate content. More precisely: operations underlying the realization of awareness should capture regularities that can be algorithmically compressible [59 - 60]; the subjective and specific quality of content does not necessarily do so. This thought experiment divides the Hard Problem into symbiotic computational and non-computational aspects[34].

The transformation of structure-based information into process-based information [55] stems from physical processes supported by real[35] numbers. A Turing Machine cannot use random or real numbers for computation. Programs use integers[36]. A random number, somewhat oxymoronically, consists of a number that a Turing Machine "cannot decide whether it is random or not" [40]. Consciousness thus has non-computational evolutionary roots[37]. Regardless of its ability to generate regularities supporting computation, brain activity stems from analogic processes.

At this point, I argue that consciousness ultimately has a non-computational nature even if it has a computational substrate. Conceivably, programmed algorithms could recognize or represent colors, but even these advanced programs would not yield an experience such as the qualia of "redness" that would transcend the algorithmic commands themselves. No algorithmic command could compel the subjective "redness" to appear inside a computer, although algorithmic commands might identify information that, for a human being, would trigger the experience of "redness." In the absence of specific commands, if consciousness should arise in a computer, it would ultimately not do so because of the specific instructions imparted on the algorithms, but as an emergent property explained by other mechanisms. We should untangle this informally grasped concept of computability [49 - 54] from the rigorously defined concept addressed by Gödel.

In the absence of specific computational commands, computation cannot be seen as an explanation for the origin of subjectivity itself. Reggia [61] summarized the current dilemma:

> *"...no existing approach to artificial consciousness has presented a compelling demonstration of instantiated[38] consciousness in a machine, or even clear evidence that instantiated machine consciousness will eventually be possible... On the other hand, and more positively, no evidence has yet been presented (including the work surveyed in this review) that instantiated machine consciousness could not one day ultimately be possible, a view that has been expressed by others."* [40]

Subjective experiences can be considered directly due to physical states of the brain or secondary to the functionality of these states, irrespective of the actual physical structure originating them. Functional states are all about what they do

[55], rather than the structures underlying them. Functionalism is compatible with a computational explanation of the Hard Problem and the possibility of instantiating phenomenal consciousness in non-biological material:

> *"It is clear, then, that proper scientific theories of consciousness are those that specify which functions are necessary for consciousness to arise. A true scientific theory will say how functions such as attention, working memory and decision making interact and come together to form a conscious experience. Any such theory will need to have clear and testable predictions that can in principle be verified or falsified. Most importantly, such theories will not claim that consciousness is a unique brain state that occurs independently of function; instead, the focus will be placed on the functions themselves and how they interact and come together to form consciousness"* [56].

The computability issue thus, provides a pivot for understanding the origin of consciousness. A non-computable consciousness position argues for an analogic processing of information relying on the laws of physics as the origin of subjectivity, and not on algorithms such as those governing digital computing. However, such a position must also account for a mechanism of transitioning information from cellular structures into network computation, systems mapping, and brain integration.

After a certain level of abstraction, it seems that the brain starts organizing information as "consciousness-objects", such as an "odor-object" (later discussed in olfaction awareness). This process can conceivably be represented computationally, arguing in favor of a computational perspective of consciousness. However, a computational stance must provide the mechanism that makes computational abstractions conscious.

Computation organizes information, and this blueprint presupposes that information must be organized in a certain manner to convey subjectivity. Conceivably, if structured information emanates consciousness the question becomes: What quality exists in this kind of information to make it a source of subjectivity? Does it originate in the cellular matter or in the later abstractions encoded in neural networks? Each possibility requires a different view of the physical laws of the universe that allow for the existence of consciousness. I chose to artificially disentangle the realization of awareness from the content of awareness to simplify the problem.

Crucially to this blueprint, the subject can be considered non-computational. There are no brain mechanisms that generate a little person who watches conscious experiences, much as a moviegoer. There is no homunculus in the

Cartesian theater. Neural correlates of consciousness refer to what the subject does.

No brain computations or mathematical descriptions of brain activities form a small someone who watches the movie of life. No neurocomputation or calculation directly creates the subject. Thus, the subject necessarily emerges non-computationally. Because there are no generative computations, I conceptualized the subject as an absence.

NOTES

[20] Gödel's theorem addresses what can be considered true in formal axiomatic mathematical systems. Gödel created a formula which states that the formula itself is unprovable in a formal mathematical system such as the one it originated from. If provable, the formula is false. Still, the formula is supposed to be provable as a matter of system consistency [50]. He demonstrated that the formal mathematical system is either inconsistent or incomplete. Regardless of which is true, it led to the conclusion that the creative processes of humans doing mathematics are not limited by formal mathematical systems, "*...which rule out the possibility that even the properties of the non-negative integers can ever be fully axiomatized.*" It poses limits to the idea of a fully algorithmically compressible human mind [50]. Readers can refer to Nagel and Newman's "Gödel's Proof" [50] for an accessible explanation of the subject and to Franzén's "Gödel Theorem" for a caution against overreaching its conclusions [52]. The reasons for Penrose's original extension of Gödel's proof into a quantum theory for consciousness are easily accessible at his popular book: "The Emperor's New Mind" [51].

[21] Classical physics does not incorporate subjectivity in mathematical or theoretical descriptions. Consciousness is not a classical observable variable. For example, there is no such concept as consciousness in classical mechanics, Newton's laws of motion, electrodynamics, thermodynamics, *etc*. In contrast, quantum mechanics allows for paradoxes of the observer as part of experiments and the object of theoretical debates.

[22] Quantum theory is an expansionist theory. Even if an explanation for consciousness is found in classical physics, the latter must remain consistent with quantum mechanics. Quantum theory demands a universal applicability to the fundamentals of nature: "*...the world really, honestly, truly is described by quantum mechanics*" [292]. *Even general relativity, so to speak the "last stronghold of classical physics", "must change, because the laws of gravitational physics must be consistent with quantum mechanics".*

[23] Quantum field theories successfully described the strong, weak, and electromagnetic interactions. Still, the framework of physics (the Standard Model plus general relativity) is not complete [292], but a hybrid between quantum and classic theories of forces. Does it have patterns and structures emerging from another *"deeper underlying theory"*? Regardless, it can be *"psychologically easy to give up on a problem, believing its solution requires new principles, when what in the end is required is simply clear thinking about the consequences of existing laws"*. I follow this advice in this book.

[24] For Alan Turing, the mathematician who broke the German Enigma code during World War II.

[25] The switching speed of silicon transistors is in the nanoseconds. The "spike" speed of neuronal messages is in average 10 spikes per second, up to about 100 spikes per second.

[26] Table lookup: a simpler data structure that indexes and fetches information, instead of using a slower and more resource intensive runtime computation or input/output operation [293].

[27] For an excellent introduction to brain computation, the reader should refer to Dana Ballard's "Brain computation as hierarchical abstraction" [40].

[28] *"It is precisely by means of the notion of 'operation' that it is possible to identify what at the same time belongs to the phenomenal conscious level and the neurophysiological level of brain activity organization, and what mediates them ...the material neurophysiological organization that characterizes the brain and the informational order that characterizes the phenomenal mind necessarily involve such events as operations at their cores. Furthermore, an operation can be at the same time viewed as an act (a process) and an object (a thing). This is so for the brain functional architectonics and for the mind phenomenal architecture's features, where brain/phenomenal patterns are both the processes and things"* [54].

[29] *"An EEG signal is composed from natural frequencies (oscillatory activity), which are traditionally divided into delta (1-3Hz), theta (4-6Hz), alpha (7-13Hz), beta (15-25Hz) and gamma (35-45Hz) frequency bands. ...Oscillatory activity of neuronal assemblies reflected in the characteristic EEG rhythms constitutes a mechanism by which the brain can regulate changes of a state in selected neuronal networks to cause qualitative transitions between modes of information processing ...fast oscillatory activity during sleep facilitates information flow from the hippocampus to the neocortex, while theta rhythm supports information transfer in the opposite direction: from the neocortex to the hippocampus. During wakefulness low-band alpha mainly related to the subject global attention readiness, whereas high-band alpha reflects the engagement of specific neural channels for the elaboration of sensorimotor or semantic information ...beta oscillations are able to synchronize neural populations over long conduction delays and might be suitable for the functional coupling of remotely distributed brain regions"* [54].

[30] Their approach allows for the assessment of mesolevel cortical interactions among transient neuronal assemblies, including a mechanism for the assessment of assembly size (by average EEG amplitude in μV) and functional lifespan (by the average length of EEG segments in milliseconds). They attempt to identify patterns in the electromagnetic field, named Operational Modules (OM), which represent a higher level of abstractness. Such analysis reveals an operational space-time within the brain internal physical space, which remains supposedly isomorphic in face of the phenomenal space-time.

[31] *"Formally 'operation' stands for the process (or series of acts/functions) that applied to an operand, yield a transform and is limited in time; and can be broadly defined as the state of being in effect. This is so regardless of whether the process is conceptual/phenomenal or physical/biological. In fact, everything that can be represented by a process is an operation. Such point of view provides a basis for the discussion of the relative complexity of operations, where there is always a more complex operation/operational act that subsumes the simpler ones. These simpler operations may also have their own complex architectures"* [54].

[32] Primitives: the simplest elements in a programming language.

[33] Symbol: a thing that represents or stands for something else.
[34] And it is indeed a preposterous thought experiment because qualia already imply the content of subjective experiences. I am carving out a separate algorithmically compressible step for the realization of these experiences. Nonetheless, the thought experiment will prove fruitful. You may consider reviewing the chapter on the "problem of records" at the end of the book. The "problem of records" chapter articulates a question requiring in my view a physical, not algorithmically compressible explanation.
[35] Real numbers represent a quantity along a line.
[36] Integer: a number that can be written without a fractional component, such as 3, 75, 0, -231.
[37] For example, the integration of a spatially mobile body with a movement-stabilized spatial environment through a shared coordinate system that occurs in the mesodiencephalon of vertebrates is considered analogic, not digital (the issue is reviewed later at the chapter summarizing the neurology of consciousness).
[38] Instantiated consciousness: the extent to which a biological or artificial system actually experiences phenomenal consciousness, qualia, and has subjective experiences. Simulated consciousness: the capture of some aspect of consciousness or its neural/behavioral correlates in a computational model [61].

CHAPTER 4

A Brief Criticism of Theories

Abstract: Two of the most popular theories that address the Hard Problem suggest that the minimal unit of consciousness could emerge from subcellular structures. Both theories neglect the importance of the subject and its agency. Both disregard the basic phenomenological definition of consciousness: there must be someone for someone to be conscious. According to the Integrated Information Theory (IIT), consciousness emanates from integrated information. The axioms at the basis of the IIT speak to the phenomenology of consciousness: consciousness exists, it is structured, it is informative, it is irreducible and it is unique or exclusive. However, the phenomenology of consciousness can only spring from the subject-object dialectic. Thus, the subject must exist a priori, as a premise, not as a conclusion. Integrated information does not provide a general cause of consciousness, but something implicit to it. The main IIT hypothesis is hopelessly flawed. A second popular theory, the Orchestrated Objective Reduction of quantum states postulates that a proto-element of consciousness accompanies any quantum reduction process. It anthropomorphizes quantum information.

Keywords: Anthropomorphic Bias, Agency and Consciousness, Consciousness in a Photodiode, Double-Slit Experiments, Effective Information, Integrated Information Introduction, IIT, IIT Axioms, IIT Fallacy, Minimal Unit of Consciousness, Missing Subject, Orch OR, Orchestrated Objective Reduction Introduction, Panpsychism, Phi-Consciousness, Proto-Consciousness, Quantum Consciousness, Quantum Gravity, Single-Celled Organisms, Subject-Object Dialectic.

Currently, work on the Hard Problem of consciousness follows two distinct approaches[39]: The Integrated Information Theory (IIT) [42] and the Orchestrated Objective Reduction Theory (Orch OR) [43]. I briefly review and criticize both theories below[40].

THE CIRCULAR REASONING OF THE INTEGRATED INFORMATION THEORY

The IIT constitutes a mathematized thought experiment.

Paulo J. Negro

According to the IIT, consciousness is integrated information, and conscious experiences are a fundamental aspect of reality. Conscious experiences are *"reduced and identical to a specific type of information"* integrated information [42]. Or, stated differently, Integrated Information emanates consciousness.

The theory characterizes integrated information by two measures [42,56]: phi-consciousness, a measure of intensity or amount of consciousness; and the nature of informational relationships that represent it[41]. Integrated information is not simply considered important or necessary for consciousness. It is consciousness in itself.

The IIT is fundamentally based on a few axioms *"intuitively"* associated with the phenomenology of consciousness:

a. Consciousness exists;
b. It is structured;
c. It is informative;
d. It is irreducible;
e. It is unique or exclusive [42].

It subsequently mathematized the thought experiment, created algorithms, calculated its integrated information theorem, and applied informational relationships in a conceptual mathematical *"qualia"* space. As a result, the IIT concludes that integrated information lies at the origin of consciousness.

Information is "classically" defined as a reduction of uncertainty. Indeed, according to the IIT, consciousness bears a direct relation to the exclusion of perceptual possibilities by a system. Cerullo pointed out that this "exclusion of information" is a central principle for the calculation of integrated information in the theory [57,58].

The theory leads to strange, however, logical conclusion. According to the IIT, a photodiode[42] that can rule out one state (light in a room) reduces uncertainty and therefore, experiences a very small amount of consciousness. A gigantic gap separates the phenomenology of consciousness in a human being and the "intuition" that a photodiode is, at some level, conscious. As presented above, large complex systems must rely on a hierarchical organization. Apart from the tremendous violation of computational abstraction hierarchies inherent to the comparison between a photodiode and human brain function, does this intuition make sense?

THE PROBLEM LIES IN THE IIT AXIOMS

A subject exists a priori in all the IIT axioms. Object and subject are dialectically bound or, in other words, the subject–object dialectic intrinsically participates in the phenomenology of consciousness. IIT phenomenological reasoning applies to subjects who are already conscious[43]. Therefore, integrated information cannot be generalized as the <u>cause</u> of their consciousness, but only as something implicit to consciousness.

In other words, because the phenomenology of consciousness reveals conscious experiences as integrated experiences, the IIT concludes that the integration is the cause of consciousness. The subject constitutes the premise, not the conclusion. As an explanation for the cause of consciousness, the IIT is hopelessly flawed. The Earth revolves around the sun, not the other way around.

The circular reasoning brings about the following problem. As the IIT does not acknowledge the pre-existence of a subject on its phenomenological axioms[44], it predicts the existence of subjects everywhere it calculates the existence of "consciousness". It is cryptically, a mathematized form of panpsychism. The IIT takes the ubiquitous existence of phi-consciousness in integrated information as evidence that organized information is in itself conscious. Therefore, I prefer to name the proposal for its logical conclusion: the phi-subject. The consequence of the central flaw in IIT can be summarized as:

Axiomatic absence of subject → Universal phi-subject

QUANTUM INFORMATION, ANTHROPOMORPHIZED IN THE ORCH OR THEORY

In the "double-slit" experiment a beam of light crosses two slits in metal plaque before reaching a screen. Because light behaves as a wave, when it reaches the screen, it forms a specific pattern with multiple bands that either reinforce or cancel each other, depending on the interference among wave peaks and troughs. The same pattern of interference is found when quantum particles, such as photons[45], are emitted one by one, apparently as an indication that each particle crosses the two slits simultaneously, interfering with itself. The interference is eliminated if the particle's path is observed, even when there is no disturbance of its motion. Assessing whether the particle crosses the slits changes the outcome of the experiment. The observer creates an effect, which provides the basis for the idea that consciousness affects quantum phenomena.

The Orchestrated Objective Reduction of quantum states, or Orch OR, postulates that a minuscule proto-element of consciousness accompanies quantum reduction processes. Measurements in quantum mechanics involve the transition of quantum states, mathematically described by quantum waves, to observable variables, such as position and momentum. One of the interpretations of the double-slit experiment is that observation forces a quantum particle to take a position and momentum. The cloud of possible routes (or probabilities distribution) of a quantum "particle" collapses in a single well-defined state. The transition is thought to occur when the wave function collapses from superposition states into a single observable state[46].

Hameroff and Penrose believe that the quantum state reduction emanates consciousness [43]. Thus, human consciousness emerges from the organization of *"biologically orchestrated coherent quantum processes"* in the brain. According to these authors, coherent quantum processes occur in collections of microtubules[47] [62] inside neurons as a mechanism correlating and regulating neuronal synaptic and membrane activity[48].

Theory critics point to Penrose's overreach in the interpretation of Gödel's theorems - an interpretation, as previously discussed, behind his search for an explanation for consciousness in quantum mechanics. Moreover, the theory makes an enormous conceptual leap by connecting consciousness to quantum processes[49]. Theoretical and practical demonstrations of quantum phenomena in the brain meet with nearly insurmountable obstacles. Current technology does not allow for experiments that could test if the human brain uses this gap for the quantum development of subjectivity.

While ample evidence lends support to computational models of brain function, the same does not occur with the Orch OR theory. Nonetheless, searching for a theory consistent with fundamental physical entities makes intuitive sense. In the absence of a computational explanation for the origin of consciousness that is based on fundamental patterns and structures, the explanatory gap can be considered paradoxically larger in computational theories despite the difficulties in demonstrating Orch OR hypotheses.

I would not intend to discuss the physics underlying the Orch OR Diósi-Penrose quantum gravity scheme of objective reduction in this text. I find it sufficient to say that Hameroff and Penrose attribution of consciousness to quantum information has intrinsic problems [43]. They anthropomorphize quantum information or, perhaps more accurately, "consciousmorphize" the wave function collapse.

Quantum information is neither conscious nor nonconscious. Dichotomies arise after state reduction[50]. Although not measurable, Orch OR thinks of proto-consciousness as arising from quantum reduction, *i.e.* having the same origin as a classical measurable variable, such as position and momentum. Quantum information before quantum reduction cannot be "consciousness" the same way that it cannot be position or momentum. As we will see later, this is a problem arising from the bottom-up approach of the Orch OR theory. Qualia cannot be properly explained by an exclusive bottom-up approach.

Furthermore, quantum processes in microtubules regulate neuronal activity, but human consciousness emerges in superior levels of brain activity, most likely requiring actions in a global neuronal workspace. How would traces of proto-consciousness move along the brain abstraction hierarchy? Orch OR does not provide an answer for the computational problem of consciousness.

THE ABSENT SUBJECT

According to the IIT and the Orch OR, a minimal unit of consciousness is smaller than a single-celled organism. The IIT describes a measure of phi-consciousness as a ubiquitous aspect of integrated information. The Orch OR theory portrays proto-consciousness as a universal phenomenon.

Neither theory stresses the importance of the subject, and its agency. Both theories concept of a minimal unit of consciousness disregards the basic phenomenological definition of consciousness: there must be someone for someone to be conscious. This someone must be doing something to realize this information. The subject, and its actions are central to the alternative explanation that constitutes this blueprint.

NOTES

[39] I review the Global Neuronal Workspace (GNW) and the Adaptive Resonance (ART) theories in the brain macrodynamics section of this book.

[40] I did not pursue extensive reviews due to space. At any rate, the problems at their foundations are straightforward. Readers should go beyond my brief summaries and look directly at the references for an expansive introduction to these theories. A good introduction of Giulio Tononi's IIT can be found at "Consciousness as integrated information – a Provisional manifesto" [42]. The reader should refer to Stuart Hammeroff and Roger Penrose's "Consciousness in the universe" for an introduction to the Orch OR [43].

[41] These informational relationships are depicted by a mathematical construct named *"quale"*, which is settled in a mathematical *"qualia space"*. The informational relationships characterize the nature of a specific subjective experience.

[42] A photodiode is a semiconductor device that converts light into electrical current.

[43] The phenomenology of consciousness (that consciousness exists, it is structured, informative, irreducible and exclusive) presupposes a person already existing and conscious. Phenomenology studies how things appear to our consciousness, not that things appearance to our consciousness is the original cause of consciousness. The cause of consciousness is the person who is conscious, not the objects that the person is conscious about.

[44] Axiom or postulate: a statement that it is taken to be true as a premise or starting point for further reasoning or arguments [294].

[45] Photon: an elementary particle representing a quantum of light or other electromagnetic radiation.

[46] I am referring to the superposition of several eigenstates.

[47] Microtubules are tubular polymers that form part of a cell's physical frame. They maintain cell structure and participate in intracellular transport, among other roles.

[48] Their theory leans on quantum gravity. It presupposes a fundamental association between quantum state reduction and the experience of consciousness. To this end, Orch OR invokes a universal *"proto-consciousness"* that, once biologically organized, would give rise to human consciousness.

[49] In Penrose's words [295], an *"inconsistency"* exists between the Schrödinger equation and the making of a measurement in quantum mechanics. (In quantum mechanics particles are described by wave functions that evolve according to the Schrödinger equation.) This perceived conceptual gap led Penrose to postulate a noncomputational system underlying human reasoning.

[50] Conceivably, state reduction due to the observer or measurement effect would be associated with consciousness whereas the elimination of quantum effects due to the interaction with the environment in a process called decoherence would not.

CHAPTER 5

Expectation Without a Spectator

Abstract: Cellular organization provides the fundamental platform for exchanges between life's integrated processes and the environment. Cellular processes convey adaptive information about the environment. Time-organized processes and information contained within living organisms are a source of fitness. The energy flow between single cells and the world forms a communication channel indissociably made of organic matter and information. This flow occurs under the informational principles of natural selection. However, the information built up into biological systems needs to become information informing about itself to transcend across levels of abstraction, such as described in Hofstadter's strange loops. A working car engine expresses the "processual" expectation of its human designers: gas flows, then ignites, pistons move and so forth. The engine has no awareness of these expectations. The expectations are implicit to the designed physical structure. The processes that mediate physical and informational relationships of living organisms also have expectations built into them. Such expectations result from a selective natural history based on evolutionary constraints.

Keywords: Car Engine Expectations, Cellular Inference, Consciousness Original Platform, Evolutionary Constraints, Emergence of Agency, Evolutionary Fitness, Hofstadter Strange Loop, Intelligent Design, Informational Relationships, Life and Consciousness, Matter and Information as Single Entity, Natural Selection, Origin of the Subject, Original Integration of Information, Processual Expectations, Phenomenology of Consciousness, Signaling, Single-Celled Organisms, Subject-World Dialectic.

INFORMATION IS MADE POSSIBLE BY EXCHANGES BETWEEN THE ORGANISM AND ITS ENVIRONMENT

Consciousness is the phenomenology of a world presented to a subject. Therefore, there can be no consciousness without a subject. Exchanges between cells and the environment constitute a platform that makes this dialectic possible.

Single-celled organisms express complex processes and a defined biological identity and engage in exchanges with the environment under the informational principles of natural selection. Human beings, as well as other animals with possible sentience, are constituted of cells. Without engaging in speculation about

Paulo J. Negro

the limits or the definition of life, one can define cellular organization as the fundamental and universal platform of life for the original exchanges between the integrated processes of an organism and its environment[51]. The energy flow organizing the subject–world bond of cellular life makes organic matter and information an irreducible single entity.

Other authors have traced consciousness to cellular or subcellular origins [39,42]. I will show how a variation of Hofstadter's strange loop can describe an operation to convert information that is implicit to cellular systems biology into information of consequence for consciousness [60]. In order to transcend across levels of abstraction, it must be information informing about itself. This idea embodies the origin of "traces" of consciousness deep into processes underlying basic interactions between the organism and the environment, such as the initial transduction of environmental stimuli into signals[52].

A CAR ENGINE EXPRESSES ITS DESIGNERS' EXPECTATIONS

Through natural selection, physical attributes become more or less stable in a given population from one generation to the next. In more complex organisms, time-organized processes and information about patterns become a source of fitness [55,65].

Time-organized processes emerging out of structures have "expectations", although they have no awareness of them. Here, the term "expectations" refers to potential states, which despite their physical scaffold, have a relational origin. In this sense, the following text provides a useful analogy:

A working car engine expresses the "processual" expectations[53] of its human designers. Gas enters the chamber, ignites, pistons move and so forth. A working car engine has no awareness of these expectations. The information representing the designers' expectations of actual thermodynamic[54] processes is embodied in the physical structure of the car engine. Conversely, a living cell expresses "processual" expectations that emerged from eons of natural selection.

LIVING BEINGS EXPRESS EMERGING PROCESSES EXPECTATIONS

This is not an explanation based on intelligent design. In fact, I envision something radically opposite: no design, only emergence. Processual expectations built into cellular processes emerged without a design and out of nothing of their own kind.

Natural selection takes place in the physical world, but in living organisms, it leads to the emergence of physical processes enacting informational relationships

[55]. Evolution does not willingly select for anything. Over generations, its processes eliminate the unfit, without design or agency. The "expectations" expressed by physical/informational relationships of living organisms result from a selective history based on these evolutionary constraints [55].

Variability, selection, and evolutionary amplification can be approached from the perspective of realization of information across potential states. Phenotypic variability iterations with the environment can be ultimately realized as genotypic expectations transmitted across generations. In addition, epigenetic changes provide a way to adapt genomic "expectations" in response to environmental input without the need for selection among generations even though they can be passed down across generations. "Predictive-error[55]" here emerges from cellular noncomputational loops.

The restrictive effects of natural selection give rise to tremendous creative processes. Natural history does not offer laws of selection, but narratives of causal chains leading to phenotypic traits[56] [37,55].

Adaptive interactions with the environment play a crucial role in the transcendence of knowledge within an adaptive hierarchy [66]. "Knowledge" at higher organization levels represents a specific case of more general knowledge at lower organization levels[57]. A phenotype contains more specific information than the genotype. The genotype allows for the production of a variety of proteins. The group of potentially available proteins differs from the group of proteins that the organism actually synthetizes to trigger the expression of a specific phenotype among all other potential phenotypes[58]. The generality-specificity relation also applies to the connectivity of the CNS, which reflects more general knowledge than the information conveyed by specific neuronal activity.

The same argument supports the existence of human brain adaptations that underlie the development of a spoken language, reading, and the emergence of a shared culture [67]. Constraints imposed by exposure to a specific language during childhood, the epigenetic imprint of writing and reading, and culture at large serve as selection mechanisms operating among preexisting brain conformational states. The relationship between potential states and actual informational states can be articulated in phenotypic selection and expression, brain epigenetic patterns, as well as in the selection of pre-existing representations to be made available for conscious appreciation.

Only 1-2% of the human genome consists of gene coding sequences. The actual number of gene coding sequences does not vary significantly between mice and humans. However, the human genome has large noncoding regions not found in the DNA of other mammals. Jean-Pierre Changeux [68] addressed the evolutionary parsimony issue in his Dynamical Nesting of Models[59] at the level of genes, transcription factors, epigenetic action on synapse formation, and long-range connectivity domains:

> *"...any level must be driven by bottom up rules from the levels underneath, it also depends top down on higher levels via* the reciprocal interlevel bridging mechanisms."

The processual expectations of living organisms result from a history of evolutionary constraints for each given species, considering both the emergence and a complex heterogeneity of causes and effects. History is not a theoretical domain, it is a description [37]. When discussing the consequences of agency among living organisms, I solely intend to observe the consequences of a fact of natural history: the emergence of agency. Integrated informational relationships so arising will have an implicit perspective: "me".

NOTES

[51] Phenomenology carefully describes things as they appear to consciousness [288]. Single-celled organisms are not phenomenally conscious. Encoding and representing conscious experiences are not the issue at this level. Information is physically depicted [70] and the cellular platform allows for its organization in a manner that is consistent with adaptive inference processes incorporating the state and the causal architecture of the environment.

[52] The variation of Hofstadter strange loop described below is based on the relationship between potential and actual states. In their open letter, Sack and colleagues, advocate the connection of cellular biological phenomena, such as membrane excitability, with the origin of sentience [39]. The disruption and restoration of cytoplasmic electrostatic equilibrium drives the neuronal activity underlying neuronal action potentials. Changes in electrostatic equilibrium caused by ionic flow across cellular membranes will modify the relationship between potential and actual cellular states, which is consistent with the strange loop variation described in this blueprint.

[53] Alas, engines do not "expect" anything in the human sense. I am using the word to refer to implicit structural information transformed in systems information.

[54] Thermodynamics: a branch of physics that studies heat and energy.

[55] Predictive-error generated by the mismatch between expectation and incoming information.

[56] The difference between natural *history* and evolution *theory* is an important concept that will be helpful when I address limitations of the Global Neuronal Workspace theory of consciousness.

[57] The simplest non-biological cybernetic systems *"provide evidence of this relation, which holds true in biological systems"* [66].

[58] Other systems provide useful analogies. For example, an electronic medical record can put together information about patients. The software can combine information from the many fields in the program. The combined information is more complex and informative

to medical care than the individual information entered into the fields by nurses and doctors. However, at the higher level, organized information can only tell about information entered in the levels below. This pattern also holds true for MRI studies. The information depicted in the image represents information collected at a lower level by the MRI scan. Human consciousness is thus seen as representing a specific form of information derived from a general form of information depicted at lower levels of system organ- ization: *"knowledge that can be instilled at a new level of organization is always limited by what the system has known prior to the process of [creation]"* [66].

[59] The dynamical nesting of models *"assumes that the brain is a nested assembly of functional structures at multiple levels of organization, from molecules to consciousness, reciprocally inter-regulated and in constant dynamic evolution, which operates intrinsically through variation-selection mechanisms yet on different times scales, from the million years of human ancestry up to 100ms of psychological operations"* [68]. Instead of one gene encoding one protein for one phenotype, gene networks encoding neuronal networks underlie cognitive functions. Gene expression is controlled by transcription factors, which are regulatory proteins binding to the gene promoter region. Transcription factors are part of intracellular and intercellular signaling pathways. Changeux describes that cooperative gene assemblies form hierarchical trees of gene expression patterns that, themselves, create higher structural levels of organization above the gene level. From the generality-specificity perspective, transcription factors convey information to genomic expectations.

CHAPTER 6

Negative Entropy and Inference in Living Organisms

Abstract: Classical information theory ties information with physical representations. Subcellular processes generate and store information, but only integrated cellular life actually transforms these processes into biologically meaningful information. Living organisms, as open systems, extract biological order from the predictably increased disorder of their environment. They transform energy into a new form of energy with higher thermodynamic value. The negative entropy results from the accumulation of information within the living system. More information translates into greater molecular complexity and functional dynamic behavior. Time-organized nonlinear processes lead to the expansion of expectations. Natural selection necessarily results in the buildup, within cellular processes, of information about environmental states and their inferred causal architecture. Such information increases the likelihood that a given organism will navigate and react more efficiently within its environment. The same holds true for brain functions that organize adaptive global behaviors. Inferential brain systems statistically minimize surprise, or "free-energy," by generating sensory samples, optimizing perception and/or improving adaptive action. Free-energy minimization intrinsically connects perception and action, supporting an early origin for top-down generative paradigms of brain function. Q consciousness-information agrees with the concept of ensemble density, *i.e.* the probability of identifying hidden causes in the environment.

Keywords: Biological Order, Bayesian Inference, Bayesian Sampler, Dissipative Systems, Environmental States, Entropy, Ensemble Density, Free-Energy, Generative Density, Hidden Causes, Information-Carrying, Information Theory, Mass-Energy Transformation, Minimization of Surprise, Natural Selection, Open Systems, Perception and Action, Predictive-Error, Singled-Celled Organisms, Statistical Physics, Thermodynamic.

CELLS TRANSFORM SUBCELLULAR ORGANIZATION IN BIOLOGICALLY MEANINGFUL INFORMATION WITH COMMUNICATIONS CHANNELS AND REFERENCE ABOUT SOMETHING

Cellular function results from time and space interactions of three types of networks: mass-energy transformation, information-carrying and signaling

Paulo J. Negro

networks[60]. These emergent cellular functions represent a complex and precise depiction of information about the organism and its interaction with the world [69].

According to the second law of thermodynamics, entropy increases over time in closed systems. Living organisms, as open systems, extract biological order from the predictably increased disorder of their environment. "Dissipative" systems play a center role in the emergent and self-organizing mechanisms of biological complexity[61] [55]. By constraining molecular organization and energy flow, dissipative systems transform energy into a new form of energy of higher thermodynamic value[62]. The result is negative entropy[63], which corresponds to an increase in information in the system. Information translates into increased molecular complexity and functional dynamic behavior.

Classical information theory ties information to physical representations [70, 71]. Subcellular processes generate and store information[64]. However, integrated cellular life[65] promoted the emergence of these biological systems. It is the cell that transforms subcellular organization with negative entropy in biologically meaningful information. This transformation occurs through the creation of communication channels and a reference about something[66].

Cells must generate information about themselves and the world to maintain a state of negative entropy. Conversely, cells improve their adaptation to the environment and optimize their own matter by generating information [55,71]. As a result, time-organized nonlinear processes[67] lead to the expansion of their own complexity and "expectations".

CELLULAR PROCESSES AND BRAIN COMPUTATIONS SHARE AN IMPLICIT PROBABILISTIC MODEL OF THE ENVIRONMENT

An enormous gap separates single-celled organisms from human consciousness. However, the same principles that organize the processes taking place in single-celled organism can be transposed to perceptual inference mechanisms of the brain.

In the cell, adaptive coupling with the environment minimizes unexpected interactions. Natural selection increases the likelihood that information built into cellular processes will incorporate information about environmental states and their inferred causal architecture. Adaptive changes increase the probability that a given organism will recognize and navigate away from a negative environment and should increase its resiliency to negative environmental changes[68].

The same is true for brain functions that orchestrate physiological activity and global, directed behaviors. From this perspective, the brain plays the part of an "inference machine"[69] [72] that uses its hierarchical computational models to predict sensory input and minimize surprise or "predictive-error". It minimizes destructive phase transitions.

The term "free-energy" is used to quantify surprise from a statistical perspective. It actually does not denote energy, but statistical probability distributions in exchanges between organisms and the environment[70]. Decreasing free-energy, in this context, means to decrease surprise and the amount of predictive-error. It minimizes the energy use or work of the organism.

Friston & Stephan defend that adaptive systems represent the state and causal architecture of the environment[71]. Such perspective appropriately fits with the idea that single-celled organisms incorporate and enact evolutionary processes. Friston & Stephan's "ensemble density" $q(\vartheta,\lambda)$ refers to the probability of selecting hidden causes ϑ in the environment. Conversely, generative density $p(\tilde{y}, \lambda|\alpha)$ refers to the generation of sensory samples[72]. Friston & Stephan's free-energy principle states that

> "*...all quantities that can change; i.e.*, that are part of the system, will change to minimize free-energy [or surprise in the exchanges]. These quantities are the internal parameters λ and action parameters."

Adaptive behaviors can be selected by optimizing perception (λ) and/or changing outward behaviors by optimizing action (α). From a Bayesian perspective:

> *"Put simply, when the free-energy [or surprise in the exchanges] is minimized, the ensemble density encoded by the system's parameters becomes an approximation to the posterior probability of the causes of its sensory input. This means that the system implicitly infers the causes of its sensory samples. Clearly, this approximation depends upon the physical structure of the system and the implicit form of the ensemble density; and how closely this matches the causal structure of the environment. Those systems that can match their internal structure to the causal structure of the environment will attain a tighter bound."*

Friston & Stephan exemplify how principles of statistical physics can be used to explain adaptive behaviors by minimization of free-energy. Moreover, free-energy minimization is intrinsically connected to the relationship between perception and action [72]. The same holds true for information-conveying structures and processes of single-celled organisms, and information encoded in the neural networks in the brain hierarchical systems.

These authors' perspective supports an early origin for the top-down generative paradigm of Predictive Coding across cortical regions, an early origin for an intimate motor-sensory relationship[73].

Friston & Stephan provide a self-modeling frame for the conscious-like purposeful behavior of single-celled organisms. Their *"ensemble density"* $q(\vartheta,\lambda)$, which refers to the probability of selecting hidden causes ϑ in the environment, is consistent with the concept of Q consciousness-information that I previously presented. It anchors Q consciousness-information in the principles of statistical physics. The implicit probabilistic model of the environment encoded in the structure and processes of single-celled organisms may thus share the same inferential paradigm of abstract information realized across the brain computational hierarchy in human consciousness[74]. Considering the conservative outlook of natural selection, this is not surprising.

NOTES

[60] The outcome of their interactions is the fluxome: the set of fluxes generated by cellular activities. The fluxome provides a dynamic picture of the phenotypic signature of cells.
[61] Aon and colleagues [299, 300] provide a good review on the subject.
[62] The theory of dissipative structures was developed to understand structure formation in far-from-equilibrium (open) systems [72]. Self-organization depends on stationary environments conducive to appropriate exchanges of entropy and energy. Living organisms enhance environmental coupling to prolong these exchanges of entropy and energy.
[63] Entropy indicates availability of energy for mechanical work. It is also as measure of disorder or randomness of a system, or the number of ways the system can be rearranged. Higher entropy → lower order. A dissipative system generating negative entropy uses energy to increase order and cellular organization. Configurational entropy measures randomness in the distribution of matter the same way entropy measures randomness in distribution of energy [72].
[64] Innumerable feedback loops express genomic information. As a result, cellular systems may resemble computing with DNA and membranes [296], whereas biochemical paths play the role of "data crunching" processes [298]. De la Fuente proposed the quantification of biomolecular information flow in bits [297]. Moreover, metabolic attractors can regulate catalytic enzymes that modify the efficiency of connections among self-organized multi-enzymatic complexes and store these changes [69].
[65] Cells bring together functional specialization and functional integration. In addition to a global broadcast of information, human consciousness requires a dynamic balance of integrated and differentiated networks of information in the brain [297, 298]. At a much different scale, the cellular organization of information can be compared with the integrated and differentiated networks underlying human consciousness.
[66] Emerging from the dialectic between integrated cellular processes and their environment.
[67] The dynamics of living organisms have a greater degree of nonlinearity than those of

inanimate matter [301]. For example, cellular fluxome oscillatory mechanisms are best addressed with nonlinear dynamic analyses in a systems biology context [299, 300]. Reductive explanations based on well-defined causal chains underestimate the implications of emergence for understanding these phenomena.

[68] *"By sampling or navigating the environment selectively, they keep their exchange within bounds and preserve their physical integrity. ...we have taken a [fanciful] paradigm example of a non-biological self-organizing system, namely a snowflake and endowed it with wings so that it can act on the environment. A normal snowflake will fall and encounter a phase-boundary, at which its temperature will cause it to melt. Conversely, snowflakes that maintain their altitude and regulate their temperature may survive indefinitely, with a qualitatively recognizable form. The key difference between the normal and adaptive snowflake is the ability to change their relationship with the environment and maintain thermodynamic homeostasis. Similar mechanisms can be envisaged in an evolutionary setting, wherein systems that avoid phase transitions will be selected above those that cannot. ...one can replace difficult questions about how biological systems emerge with questions about what behaviors they must exhibit to exist"* [72].

[69] Including the treatment of brain function as enacting Bayesian statistics. Bayesian inference is a statistics method in which the probability of hypotheses is updated, as more information is available. The Bayes theorem addresses a (H) hypothesis that is conditional | on new (E) evidence:

$$P(H|E) = P(E|H).P(H)/P(E)$$

in which P(H|E) is the posterior probability of (H) after (E) is available, P(E|H) is the probability of (E) given (H) before evidence is available, P(H) is the prior probability of (H). P(E) is the likelihood function of the theorem. Bayesian inference is a form of computation that combines new evidence with prior beliefs.

[70] The readers should refer to Friston & Stephan for an excellent summary of the subject. In thermodynamics, "free-energy" is defined as the amount of work extracted from a system, or the difference between the energy and entropy of such system. The thermodynamics "free-energy" and statistical "free-energy" share the same mathematical descriptions [72] and thus the somewhat awkward term to describe surprise or prediction-error. Note that this mathematical description of such exchanges means that the biological systems in these organisms make inferences about their surroundings and depend on the exchange of entropy for biological self-organization.

[71] A brief summary of the heuristic work developed by Friston & Stephan [72]:

Sensory inputs emerge from high-dimensional interactions between action and hidden causes:

$$p(\tilde{y}) \rightarrow p(\tilde{y} \mid \alpha)$$

describes environmental effects on system $\tilde{y}$ and its sensory input (or state of sensory receptors). α represents forces applied by effectors to change sensory samples. α adjusts

sensory input by interacting with the forces that cause $\tilde{y}$. α has hidden environmental causes ϑ.
Biological agents stay within phase-boundaries by controlling $\tilde{y}$ through adaptive changes in α. They minimize unlikely or surprising exchanges with the environment:

$$Q(\tilde{y}|\alpha) = \ln p(\tilde{y}|\alpha,m)$$

where m depicts the agent, phenotype or system and $p(\tilde{y} \mid \alpha,m)$ the effects in conjunction of α and m on $\tilde{y}$. "Any member of a population, whose population density $p(\tilde{y} \mid m)$ is at equilibrium, must on average increase $Q(\tilde{y} \mid \alpha)$." The conditional surprise $-\ln p(\tilde{y}|\alpha,m)$ measures the "improbability of exchange, given a particular agent and its action".

$$Q(\tilde{y}|\alpha) = \ln \int p(\tilde{y}, \vartheta|\alpha)\, d\vartheta$$

$Q(\tilde{y} \mid \alpha)$ is thought as the "adaptive value of a particular exchange". Where $p(\tilde{y},\vartheta|\alpha)$ is the joint density of environmental effects and their unknown causes.

$$F(\tilde{y}, \lambda|\alpha) = -Q(\tilde{y}|\alpha)$$

where F represents free-energy, $\tilde{y}$ is the sensory input or environmental impact, λ depicts internal states or variables, and (not shown, but used for the calculation) $q(\vartheta,\lambda)$ is the "ensemble density", or the probability density that causes ϑ would be selected from an ensemble of environments.
[72] "*The generative density factorizes into a likelihood and prior density* $p(\tilde{y} \mid \vartheta,\alpha)\, p(\vartheta)$, which specify a generative model. This means that the free-energy formulation induces a generative model for any system and an ensemble density over the causes or parameters of that model ...these forms enable the free-energy to be defined as a function, $F(\tilde{y}, \lambda|\alpha)$ of the system's sensory input and internal state".
[73] Including, at the subcortical level, the close relationship between deep-layer motor fields and superficial-layer sensory fields in the superior colliculus, as I will discuss later.
[74] As quoted by Nicolelis [207]: "*Whether or not living things are aware of it, they are embodiments of theories about the laws of Nature drawn from the part of Nature they have encountered*".

CHAPTER 7

Expectations in a Strange Loop

Abstract: Complex systems that handle information follow a hierarchical order. The realization of information in hierarchical systems occurs as feedback loops that break the hierarchical order and generate self-referential information. The violation of hierarchical order defines a paradoxical level-crossing strange loop. This pattern lies at the origin of an "I". It can be depicted as a mathematical phrase composed of *'subject'predicate* that undergoes a loop through which the predicate becomes its own subject: *'predicate'predicate*. Processual information includes present and potential information expressed as *information.expectation* that becomes realized as the strange loop *'information.expectation'information.expectation*. In fact, expectations drive this strange loop, which can thus be expressed as *'expectation.information'expectation.information*. Conscious experiences constitute running models (expectations) that simulate reality. As multiple levels of strange loops realize *expectation.information* across the hierarchical organization of the brain, consciousness becomes a higher-level abstraction that supplants and integrates other abstraction levels. Perception inversely maps or generates hidden causes from sensations. Abstractions arise from seen and unseen actions of the organism, which become the source of expectations across multiple levels of nested brain oscillations. Actions provide the basis for this generative model of consciousness. Neural correlates of consciousness do not necessarily explain the origin of consciousness but may simply describe regions implicated in its realization.

Keywords: Active Abstractions, Analogic Realization of Information, Brain Rhythms, Complex Systems, Consciousness-Information, *Expectation.Information*, Gödel's Proof, Hierarchical Structure, Hofstadter Strange Loop, *Information.Expectation*, Inverse Mapping, Level-Crossing, Neural Correlates of Consciousness, Primitives, Potential States, Processual Information, Self-Modeling, Self-Referential, Variable Q.

THE INTRODUCTION OF 'EXPECTATION.INFORMATION' EXPECTATION.INFORMATION

Biological and silicon-based complex systems that handle information bear a hierarchical structure [42]. Douglas Hofstadter extensively[75] wrote about mechanisms of self-reference in hierarchical systems as a prerequisite for the emergence of awareness of information. According to him, the system must

Paulo J. Negro

"engulf itself" in a twist, violating its hierarchical order. Hofstadter called this twist a "strange loop" [64]. Thus, he defined a strange loop as a paradoxical level-crossing feedback loop. Much like Penrose [43], Hofstadter was deeply influenced by Gödel's proof, who considered the quintessential mathematical strange loop.

In tandem with Gödel, Hofstadter depicted a strange loop as a mathematical phrase composed of

'subject'predicate

That undergoes a loop through which the predicate becomes its own subject:

'predicate'predicate

This phrase composition represents the fundamental concept behind Gödel's proof. It characterizes a strange loop as an expression of feedback breaking the hierarchical order and generating self-referential information. This event, according to Hofstadter, should be seen as the informational origin of an "I" [64].

The realization of processual information requires a self-referential loop including present and potential (absent) information. The silence of potential states also constitutes information. The processes interconnecting organism and environment can be expressed as:

information.expectation

Which represents a relationship between actual and potential processes, and the origin of this strange loop:

'information.expectation'information.expectation.

This format, originally derived from Gödel's proof, depicts processual information deriving information about itself, its own potential states and its interactions with the environment:

'information.expectation'information.expectation

≈

realization of consciousness-information

The processes of cellular life amount to the expressions of an irreducible single entity combining organic matter and information. Cellular activity involves innumerable loops of interaction, which characterize complex dynamic systems, but a formal divide between software and hardware does not exist in cells. Thus, digital algorithms cannot realize the expectations of cellular processes, although may be used to describe cellular activity. As a consequence of this noncomputable stance:

Consciousness-information ≈ analogic realization of information

As a result, the realization of “Q” does represent a variable for the “act of processing reality into subjectivity”. It represents organic matter/information realizing information about organic matter/information processes of subject–world exchanges through the laws of physics. At its roots, Q serves as a variable that turns the direct access to the reality of the physical world into an abstraction.

However, this interpretation of Hofstadter’s strange loop neither stresses the physiology of brain mesodynamics nor does it emphasize consciousness as action. A better interpretation should match the perspective of Predictive Coding (Fig. **2**), and I will later provide detailed reasons for this new structure:

‘expectation.information’expectation.information

≈

‘feedback.feedforward’feedback.feedforward

A GENERATIVE MODEL OF CONSCIOUSNESS IS BASED ON ACTIONS

Multiple levels of strange loops that realize *expectation.information* across the brain’s hierarchical organization can make consciousness an abstraction supplanting and integrating other abstraction levels. At some point, the brain most likely has the ability to process information computationally or, more precisely, at some point, computational approaches to brain function become overwhelmingly successful. This hybrid model for a consciousness of noncomputational roots and computational abstractions may provide an adequate frame for the Hard Problem. I will return to the issue later.

This generative model of consciousness, much like any such model, has a basis on action. Perception inversely maps or generates hidden causes from sensations. Abstractions arise from seen and unseen actions of organisms, which become the source of expectations across multiple levels of nested brain oscillations and become realized across hierarchical levels by the *expectation.information* version of Hofstadter’s strange loop.

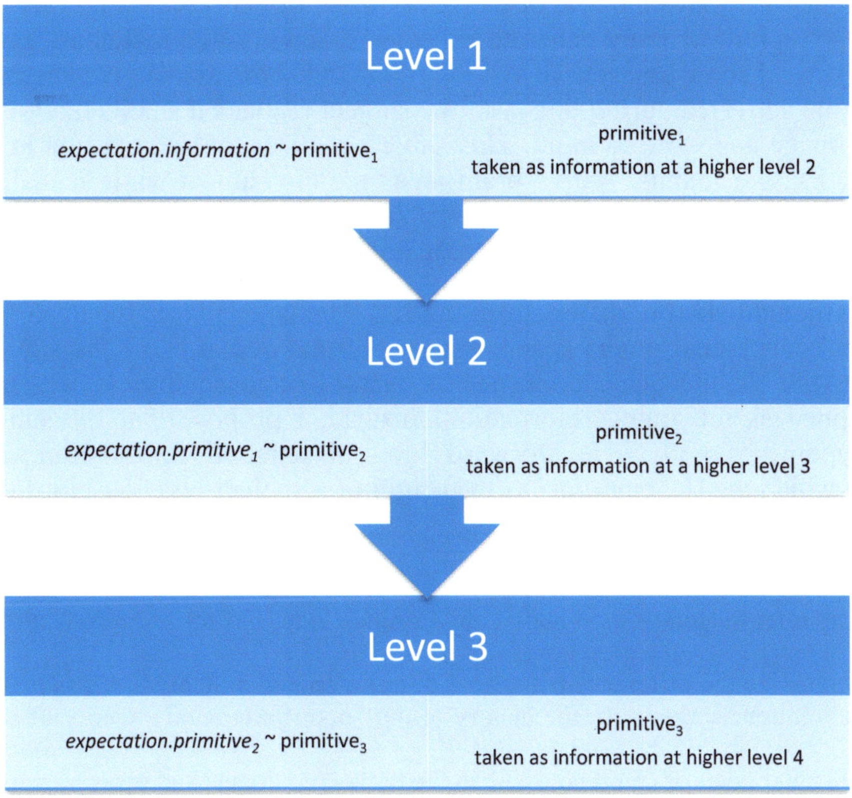

Fig. (2). Unpacking *'expectation.information'expectation.information* as a multiple level self-modeling description in which lower levels provide primitives for higher-level modeling. The strange loop reaches deeply into lower levels of organization. This model illustrates the transition between computational and noncomputational relationships of "expectation" and 'information", ultimately anchored in subcellular processes.

Among the complex interactions of the brain's sustained rhythms, it should be possible to verify errors and their variance from expectations. Across levels, information can become expectation, and *vice versa*. Conscious experiences, sustained by nested rhythms, thus behave like running models (expectations) that simulate reality.

Neural correlates of consciousness do not necessarily explain the origin of consciousness[76]. They may simply describe the areas implicated in its realization. Highlighted brain areas generate predictions and/or match these predictions with incoming information. These regions identify what the subject does. Think about some olfactory experience. A functional imaging study could show activation of the piriform cortex[77] and the right anterior orbital frontal cortex (OFC). It matters not if the right OFC plays the role of a primary olfactory cortex or a multimodal cortex. These regions actively generate patterns of activity that match an "odor-object"[78]. Now, if some associative vision region and some associative language region become activated during this task, the subject realizes it smells like roses, and brings image and word to mind. This subject realizes the information in the language and visual realms. Active brain areas are the subject while it realizes information. In other words, perception contributes to the making of the perceiver. Such interpretation avoids the cryptic encroachment by intelligent design.

This line of thought sharply differs from the IIT. The latter places the quale in a mathematical conceptual qualia space, with no formal depiction of the subject. According to the IIT, this quale has a specific quantity of consciousness, which, in turn, is defined as integrated information. Instead, I propose that the subject matches hypotheses with a feedforward loop of external information. The integration, which the IIT pinpoints as the origin of a subject-less consciousness, is the consequence of the subject's actions.

NOTES

[75] Readers can refer to Hofstadter's book I am a strange loop [64] for his introduction of the subject.

[76] I again emphasize the difference between a description and a theory, such as natural history and evolutionary theory or the history of our solar system and a theory of solar systems.

[77] Cortex: the outer layer of neuronal tissue in the brains of humans and other mammals. A large medial fissure divides it into left and right hemispheres. Subcortical regions, located under the cortex, include evolutionarily older structures.

[78] I describe the concept and evidence of odor-objects in the section reviewing the olfactory system.

CHAPTER 8

In Search of Meaning

Abstract: Natural selection acts on emergent informational relationships that govern the physical processes in living organisms. As complexity increases, and processes affect other processes under the constraints imposed by natural selection, referential information can become relational and abstract. Single-celled organisms do not experience meaning. They have no consciousness. Nothing appears to their consciousness in a phenomenological sense. The information that accumulates as these organisms dialogue with the environment has a referential relationship with the processes of natural selection. Classical Shannon's information theory (SIT) addresses conditions for something to convey information, but not the meaning of information. Meaning emerges from decoding of information, which, in turn, can only occur in reference to something external to Shannon's information. Natural selection may provide meaning because the processes that living organisms carry out become or contain metainformation about natural selection itself. Meaning and abstract awareness share the same fundamental mechanism. These considerations suggest that the loop *'expectation.information'expectation.information* realizes meaning and provides a path towards the solution of the "aboutness" problem. The organization of information as abstract and referential may allow for partial decoupling from underlying thermodynamics.

Keywords: Aboutness Problem, Brain Metabolism, Decoding Information, Entropy and Uncertainty, Evolutionary Constraints, Energy Budget, Evolutionary Metainformation, *Expectation.Information*, Information About Something, Origin of Agency, Referential Function, Realization of Meaning, Phenotype, Shannon's Information Theory (SIT), Thermodynamic Decoupling.

SHANNON'S INFORMATION CAN BE REALIZED AS INFORMATION ABOUT SOMETHING

A solution to the problem of the origin of subjectivity must take into account the origin of meaning. The IIT and Orch OR fail to directly address this crucial challenge.

Shannon's Information Theory (SIT) addresses conditions for something to have the potential to convey information, but not a frame of reference for information [70]. It does not colloquially refer to "information" as a term that denotes something meaningful.

Paulo J. Negro

Meaning becomes possible when "information", understood as reduced entropy and uncertainty, acquires a referential function. For example, not receiving "information" provides information within the expectations of a referential function[79]. The absence of a war declaration informs of ongoing peace. The absence of a tax refund check informs the tax payer that something has happened during the process of tax refund, including different possibilities such as lost mail, incorrectly filled tax return forms, tax agency mistakes or identity theft. These examples apply the colloquial use of the term information which is information "about" something [55].

Information, as defined by SIT, can only convey meaning after being decoded. However, decoding can only occur in reference to something external to Shannon's information. In other words, meaning can only exist in a referential context. It must be meaning about something.

Deacon called it the *"aboutness"* problem [55]. The *"aboutness"* problem underlies the failure of computational theories to explain the origin of consciousness: *"computation assigns referential values to physical processes and exacts physical manipulation of process variables"*. The Orch OR theory contains the same limitation. Mapping of information conveyed by quantum phenomena does not solve the problem that quantum phenomena do not provide meaning. The problem of absent content remains unchallenged.

Evolution has no inherent goals. However, evolutionary history resulted in organisms that functionally aim towards survival and reproduction [37]. The emergent informational relationships governing physical processes in living organisms allow natural selection to have a surrogate action on them, in spite of its absence of agency. Natural selection has no agency, but organisms do. There are no goals in natural selection, but organisms act towards the goals of survival and reproduction[80].

As complexity increases and processes influence processes under the natural selection framework of constraints, it becomes possible for referential information to become relational and abstract [55]. Abstract/relational information emerges from physical relationships, in contrast to computational models that emerge from abstract algorithms. The physical background generates metaphysical relationships.

As reviewed by Deacon, the constraining influence of natural selection can transform noise into signal and shape the processes and structures of the system facing its influence [55]. For example, mutations, that can be considered noise in the genetic code, can become signal over generations if they yield phenotypic adaptations. Adaptations generate a referential context and functional characteristics that enhance the connection to their own context. Note the implicit strange loop of "adaptations enhancing connections to their own context" that characterize evolution.

Single-celled organisms do not experience meaning. They have no consciousness. Thus, nothing "appears" to their consciousness in a phenomenological sense. Information structured from the organism–environment dialectic is a referential relationship "about" the processes[81] of natural selection. At a fundamental level, natural selection may provide meaning because these physical processes become or convey metainformation about natural selection. The fact that cellular processes that enhance adaptive coupling to the environment are selected, by itself, conveys information about natural selection that goes beyond the information contained in what is selected in a specific situation. The referential relationship to natural selection as the origin of meaning is abstract, relational, a form of meta-information beyond specific genotypes or phenotypes.

These considerations allow for a different understanding of

'expectation.information'expectation.information

Where processual expectation is a necessary aspect of the referential context and indispensable for the realization of meaning in its most abstract form. The "aboutness problem" may be solved as:

'expectation.information'expectation.information ≈ realization of meaning

In other words, abstract awareness and meaning share the same fundamental mechanism. The human ability to find meaning relates to the conclusion reached by Penrose from Gödel's proof: that the "understanding" of the human mind transcends algorithms. However, quantum phenomena do not necessarily provide the basis for this ability. "Understanding" emerges from a relational order of non-Shannon information about long-lasting evolutionary processes. Human meaning and life itself share the same evolutionary origin, albeit with different orders of organization. Thus, this blueprint causally connects life and consciousness, and avoids the anthropic characteristics of the IIT and the Orch OR theory.

Regardless of mechanism, the human brain certainly processes non-Shannon's information. Non-Shannon's information is referential information that indicates meaning. Referential information may partially decouple from the underlying thermodynamics. The thermodynamics of Shannon's information does not apply to non-Shannon's information, as often observed outside the brain: a declaration of war (non-Shannon's information) causes energy exchanges in excess of the energy necessary for its electronic transmission (Shannon's information) [55]. Of course, the referential context of a declaration of war depends on the referential context of human minds and the organization of information as meaningful information rather than information as defined by SIT.

The brain operates under budget constraints[82] [73, 74]. A narrow range of energy use indicates that the brain processes information with different degrees of meaning complexity with a similar level of thermodynamic coupling. Energy use depends mostly on Shannon's information, such as the recruitment of different brain regions by attention, use of working memory, retrieval of hippocampal information, *etc.*, not on the aboutness of information. The metabolic costs of human brain development may explain the constraints that favor reliance on non-Shannon's information.

NOTES

[79] I recommend that readers review Deacon's excellent book "Incomplete nature: how mind emerged from matter" [55]. His ideas strongly influenced this manuscript. His book details several positions presented here, including a mechanism for abstracting information away from its conditions of origin.

[80] I am not reifying processes of natural selection but pointing out the consequences of orders of organization emerging from the agency of living beings. Because organisms are agents in their environment, from their functional perspective survival and reproduction are evolutionary goals.

[81] I do not intend to hint at a subtle intelligent design or to give agency to natural selection. I use the term "processes of natural selection" to designate the consequences of evolutionary constraints.

[82] *"Not only there is a maximum limit of firing that an ensemble of neurons can reach, but the global ensemble firing rate tends to stay constant, hovering around a mean, due to a variety of compensatory mechanisms that create stable equilibrium. If single or multiple cortical neurons increase their firing rate instantaneously, an equivalent mirror image reduction in firing is soon produced by other members of the neuronal ensemble so that the overall energy budget of the brain stays constant over the long run."* [207] This may be confirmed by functional neuroimaging studies showing a parallel of increased and decreased metabolic activity in brain regions across different tasks posed by different study paradigms.

CHAPTER 9

Supplanting the Integrated Information Theory

Abstract: Natural selection does not start *de novo* with each new species; it uses the information contained in organisms that have come earlier in evolutionary history. Evolution is conservative. A self-generated consciousness can only emerge as the product of information realized out of the exchanges between the organism and the world. Cells provide the original platform that first organized this kind of integrated information exchanges. They are not conscious, but their systems biology is a form of knowledge. Cells are the original knowing without a knower. Human consciousness is also abstract, and one cannot directly measure it. Its abstract nature is a description, not a speculation. The IIT relies on the concept of effective information, which represents the entropy of the actual repertoire of a system in relation to its potential repertoire. This aspect of the theory agrees with the notion of *information.expectation* and renders this blueprint consistent with the empirical support of the IIT. However, the IIT does not address the role of the subject in the realization of *information.expectation*. The realization of information is an active process. Neurobiology provides no evidence for the production of information by unactivatable systems.

Keywords: Action-Based Consciousness, Complex Emerging Systems, Consciousness in Inactive Systems, Consciousness Detachment from the Subject, Conservative Natural Selection, Evolutionary Origin of the Subject, Effective Information Calculation, Empirical Support, Entropy, Information Single-Celled Organisms, *Information.Expectation*, IIT Fallacy, Integrated Information, Intelligent Design, Minimally Conscious Photodiode, Overlap with the IIT, Realization of Relational Information, System Repertoire.

HUMAN SUBJECTIVITY IS ABSTRACT AND CANNOT BE MEASURED DIRECTLY

This blueprint rests on simple ideas. A self-generated consciousness can only emerge as the product of information realized out of the exchanges between the organism and the world. In this sense, cells represent the original platform that first organized this kind of integrated information exchanges. Cells are not conscious. However, their systems biology is a form of knowledge. Cells are the original "knowing" without a knower.

Paulo J. Negro

Human consciousness is abstract and cannot be measured directly. The abstract nature of human consciousness is a description, rather than a speculation. The brain is as empty of consciousness as a cell is empty of consciousness-information, notwithstanding our vivid phenomenal consciousness and selfhood experience. No conceptual leap is needed here, just inferences about the causal structure of the world at different levels of organization.

Natural selection does not start *de novo* with each new species; it builds on organisms that have come before [75]. Evolution is conservative. The management of biological systems' information represents an adaptive feature of singled-cell organisms. Therefore, one might expect to find such informational structures at the basis of the highly complex systems that resulted in the emergence of awareness.

Traditionally, physiology addresses the interaction among specific compartments within an organism. For example, in the process of glutamatergic[83] synaptic transmission, the pre-synaptic neuron releases glutamate in the synapse and stimulates post-synaptic sites. As a result, the post-synaptic neuron produces and release endocannabinoids[84] into the synaptic cleft[85]. These neurotransmitters stimulate the pre-synaptic membrane CB1 endocannabinoid receptor[86] suppressing further releases of glutamate [76]. This negative feedback loop represents but one of the many loops in the central nervous system. Scientific knowledge depends on understanding each part of this process: knowing the parts means knowing the loop.

However, the information that emerges from processes involving large molecular ensembles has value for the overall synaptic regulation. Managing the interactions among complex emerging systems at the cellular and intercellular levels yields an adaptive premium. The relational information that emerges from complex systems does not exist physically beyond the parts that make it. In parallel, the word "consciousness" applies to the non-physical aspects of the human experience.

In the dynamics of cellular systems, I have chosen the words "knowing" and "expectations". In particular, I reviewed the role of local networks and mechanisms of local "awareness" as intermediates between cellular information and the global sharing of information characteristic of human consciousness. I chose examples of such networks in areas supporting specific awarenesses[87], either because they involved relatively simple processes or because they had heuristic value: awareness of location, awareness of time, interoception, and olfactory awareness.

The review of studies on local circuitry will be presented next, followed by a review of theories on the global sharing of information as the basis for human consciousness. At that point, the reason I support a subject-centric theory of consciousness should become clear.

THE IIT IS A CRYPTIC FORM OF INTELLIGENT DESIGN

Before going through the neurobiology, I will explain why this blueprint should be consistent with the empirical evidence supporting the Integrated Information Theory. According to the IIT, the entropy of a system's actual repertoire relative to its potential repertoire represents the effective information generated by such a system[88] [77]. In other words, the IIT provides a mathematization of *information.expectation* without addressing the role of a subject in the realization of *information.expectation.*

Based on the concept of effective information, the IIT allows for the calculation of integrated information by comparing the "actual" repertoire generated by a system as a whole with the combined actual repertoires generated independently by its parts[89]. Subsequently, the IIT mathematizes its informational structure, generating a *"maximum integrated set of elements and a maximally irreducible conceptual structure"*[90] which is reported as describing the informational qualities of qualia.

I already addressed the central flaw in the IIT axioms, which indicates that phi-consciousness cannot represent a "cause" for consciousness. Instead, phi-consciousness provides a surrogate marker for the "traces" of consciousness occurring across brain hierarchies.

Still in accordance with the IIT, consciousness emanates from integrated information: a minimum degree of consciousness is portrayed to exist even in inactive and in simple systems, such as a *"minimally conscious photodiode"*. However, even this photodiode is presented as activatable. Neurobiology provides no evidence for the production of information by systems that cannot be activated, although the potential states of systems participate in the realization of information. In other words, dead biological systems are not the same as inactive systems. They have no potential information because they have no potential to become active. The calculation of the effective information in a dead system or tissue is the mathematization of a truism.

Proponents of the IIT commit the same mistake incurred by creationists who did not believe evolution could account for the richness and complexity of living beings. Creationists could not fathom the mind-boggling time-span of evolution[91]. The first experiences of simple awareness occurred sometime early in the natural history of this planet, and this natural history generated consciousness as one of its byproducts. An unimaginably long chain of organisms connects the information contained in the most primitive emerging systems with human consciousness. From a natural history perspective, a subject has always existed. One should avoid falling into the intelligent design pitfalls of the artificial detachment of consciousness from its subject.

NOTES

[83] Glutamate: the most common excitatory neurotransmitter.

[84] Endocannabinoids: endogenous neurotransmitters sharing properties with the cannabinoid molecules of marijuana.

[85] Synaptic cleft: the space between neurons in which communication chemically occurs through the release of neurotransmitters.

[86] A receptor receives a chemical signal from outside the cell and triggers a physiological response.

[87] Keep in mind that the creation of this variable "Q" to designate inference processes (and consciousness-information) allowed me to explore perplexing concepts, such as the idea of local awareness.

[88] "*Information:Consider [a system representing] a binary photodiode. The system has a mechanism such that if the sensor is on (S state = 1), the detector turns on (D state = 1), and is currently in state [11]. How much information is generated by system X, endowed with causal mechanism mech being in the particular state* $x_1=(n^1{}_1n^2{}_1)=[11]$ *at time t=1? Prior to considering its mechanism and current state, the system of two binary elements could have been in any of four possible states ([00],[01],[10],[11]) with equal probability p=1/4. This potential repertoire (or "a priori repertoire") is the maximum entropy (maxent or uniform) distribution, which entails maximum ignorance, indicated with* $X_0(maxH)$*. The mechanism and current state of the system, however, reduce uncertainty, i.e.* generate information, about the previous state of the system (at t=0). This is because only some previous states (in case, [10],[11], with equal probability p=1/2) could have led to the current system state x_1 through the mechanism X_0(mech), while the previous states [01],[00] could not (p=0). In general, mechanism and current state specify an actual repertoire (or "a posteriori repertoire") or $X_0(\text{mech},x_1)$, the probability distribution expressing how compatible previous system states are with the system's mechanism and current state. The effective information generated by the system is the entropy of the actual repertoire relative to the potential repertoire (also known as the Kullback-Leibler divergence): $ei(X_0(\text{mech}, x_1)) = H[X_0(\text{mech}, x_1)||X_0(\text{max H})]$" [77].

[89] Integrated information is the entropy of the actual repertoire of the system relative to the repertoire of its minimal parts (*i.e.*, the product of a decomposition of the system that leaves the least information unaccounted for, called minimum information partition [77].

[90] This is consistent with a small-world architecture. A small world architecture supports the maximum complexity of connectivity giving rise to a scale-free system that obeys a power law. The IIT indicates that systems capable of generating the most integrated information and the best discrimination among informational states should be empirically associated with brain structures related to the high complexity of human consciousness.

[91] "*...from so simple a beginning endless forms most beautiful have been, and are being evolved*" [302].

CHAPTER 10

Awareness of Location

Abstract: Rodent studies of navigation across single- and two-dimensional environments show patterns of hippocampal pyramidal cell activation characterized by a "place field" in the format of a cone and a highest firing "place cell" neuron in its center. As the animal moves across the environment, the field of neuronal activation concomitantly changes in the hippocampus. The animal realizes knowledge about its location through the relationship between a central neuronal spike and the overall activation of a place field. Awareness of location is signaled by the matching of maps represented in the hippocampus with incoming perceptual information. This pattern supports the idea of subjective experiences as abstract information that transcends specific physical location. It supports a content-independent mechanism for the realization of information into "knowledge". In two-dimensional mazes, neurons that become active at the crossroads also become tied to different routes, emerging as omnidirectional cells. Multiple episodes of common junctions can free neurons from their physical context. Emergent levels of omnidirectional assemblies have a relational origin and can become part of higher order representations. These results suggest the existence of mechanisms that abstract information from spatial and temporal perspectives. They also indicate that motor actions have a pivotal role in the generation of maps and, ultimately, that actions form the basis for the representation of the physical world and the transition of Shannon to non-Shannon "aboutness" information. Such results counter the IIT premise of awareness generation in systems that cannot be activated.

Keywords: Aboutness Problem, Abstract Maps, Action-Based Consciousness, Awareness of Location, Bottom-up Consciousness, Content-Free Qualia, Feelings of Knowing, Hippocampal Navigation Studies, Hippocampal Pyramidal Cells, Higher Order Representations, Hippocampal Theta Oscillation, IIT Inactive Systems, Omnidirectional Cells, Oscillatory Phase, Place Field, Place Cell, Realization of Meaning, Spike-Timing-Dependent Plasticity, Temporal Code, Thermodynamic Decoupling.

Awareness of location requires specific perceptual information. However, the mechanism that signals awareness of location is not the same as the specific perceptual information necessary for its realization. Instead, awareness of location arises from the matching of maps, represented in the hippocampus, with incoming

Paulo J. Negro

perceptual information[92]. Because the hippocampus is one of the best characterized networks, these experiments have heuristic value [78].

A BRIEF DESCRIPTION OF THE PHYSIOLOGY

Since 1971, studies on rodents found that hippocampal[93] neurons code for their Cartesian position, regardless of the animal's behavior and direction of movement[94]. In these studies, rodents navigate through either a single dimension on a straight line or through more ecologically representative two-dimensional environments. In the two-dimensional environment, measurements of individual hippocampal pyramidal cells display a pattern of activation characterized by a "place field" in the format of a cone, with the highest firing "place cell" neuron in its center. In each location, only a fraction of the hippocampal pyramidal cells[95] fire, most remain silent. As the animal moves across this environment, the field of neuronal activation concomitantly changes (Fig. **3**).

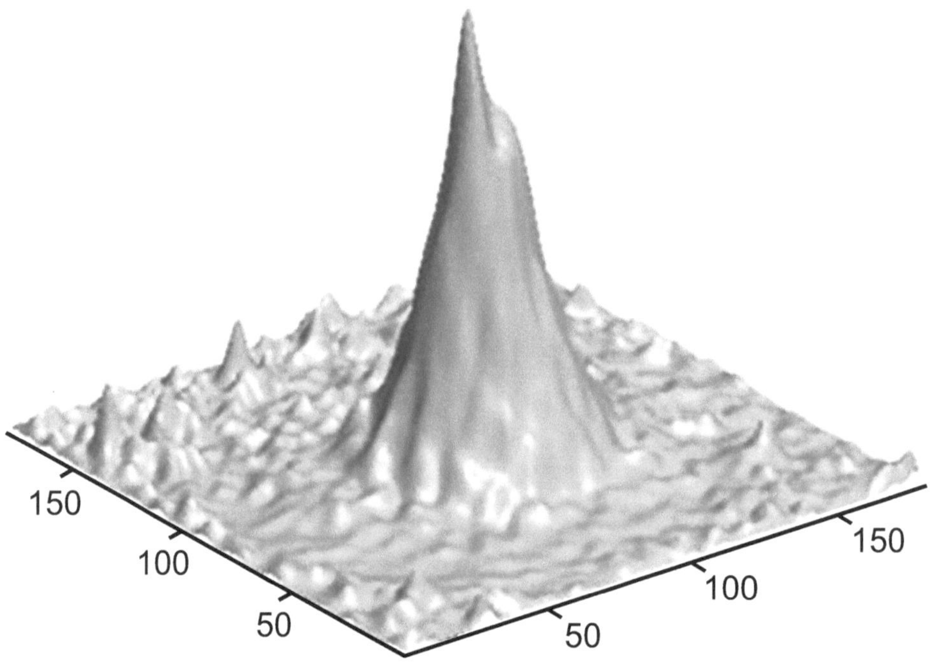

Fig. (3). Omnidirectional place cell cone: positional representation by a CA1 place cell in a two-dimensional environment: X- and Z-axes show distances in centimeters; Y-axis shows the neuron's firing rate. Its shape is independent of velocity, acceleration, trajectory, but the frequency of discharge (height) is. Reprinted with permission from Samsonovich and McNaughton and the Society for Neuroscience [79].

Single dimension experiments yield this pattern more clearly. In this case, the animal walks through a track while the activity of specific pyramidal cells is measured. Neurons originally in the periphery of the activation field increase their firing as the animal progresses. Neurons close to the activation center decrease their firing as the animal moves forward, and these neurons represent the tail end of a new place field configuration, which comes to characterize the new position in the track.

The animal knowledge[96] of its location is not given by any specific neuron, but from the relationship between a central spike of neuronal activity and an overall activation of a place field. Neuronal firing generates effective information in the context of an active neuronal assembly representing potential alternatives[97]. It is a local form of *information.expectation*.

The information is further integrated in the context of the overall behavior and body position. For example, neurons outside the hippocampus code for head direction[98] and entorhinal cells[99] fire in a hexagonal grid pattern encoding an overall representation of space. While place cells are likely candidates for the neural implementation of Tolmanian[100] cognitive maps, grid cells[101] convey spatial selectivity to those place cells located downstream in the hippocampus[102][80, 81].

The subjective experience of being in a place is abstract information transcending specific physical location. It is the relationship between place cell and place field that signal this kind of knowing. This information pertains to the fine-grained encoding of location by the hippocampus. Neurons at the perirhinal cortex[103] further process and discriminate complex environmental information, including the integration of separate items in one representation [80, 81].

The act of knowing may thus exist regardless of content or qualia. Studies of awareness of location provide a helpful approach to the Hard Problem for this exact reason. They allow for a breakdown of the problem in two complementary parts. The first addresses the mechanism underlying the production of awareness itself, and the other addresses the mechanism that generates the "content" of awareness: *information.expectation* is hereby seen as a content-independent mechanism for "knowing". We will see later that these considerations match the observations contained in the Global Neuronal Workspace theory regarding abstract states of awareness, such as feelings of knowing (FOK) in contrast with states of awareness traditionally associated with qualia, such as vision and color.

CAN QUALIA BE PRESSED DOWN THE RABBIT HOLE OF LOCAL NETWORKS?

These studies provide further evidence against the IIT. Specific neuronal networks can represent location only because the organism as a whole experiences location. The brain builds cognitive maps based on the animal's previous experiences.

There is no way to "impress" the informational correlates of a subjective experience upon specific networks without an organism. IIT phi-consciousness is content-free.

The IIT proposes a bottom-up generated consciousness without a subject. The problem must be addressed from another direction, *i.e.* an explanation for how a subjective experience is "pressed" upon specific neural networks. In other words, if qualia were something physical to be imprinted on specific neurons, this imprinting could not happen in the absence of an organism because the overall experiences of the organism are the source of qualia. While I just described a "content-free" mechanism for the realization of awareness of location, its realization cannot occur without the animal actually seeing, hearing, and smelling its environment as a global action.

IIT phi-consciousness is empty of qualia because there is no mechanism to "impress qualia" upon neural networks. The IIT presumes a generic form of consciousness dissociated from organisms. Phi-consciousness describes qualia, but it is devoid of qualia.

Qualia have no physical existence that become imprinted on neural networks. The brain cannot store qualia. Qualia refer to the subjective experiences that emerge when an organism acts within its environment, as such, qualia always happen *de novo*[104]. Nevertheless, something local must still happen to promote qualia. As we will see later, the interaction between top-down and bottom-up informational streams may offer a solution to this paradox.

REALIZATION OF INFORMATION ACROSS OMNIDIRECTIONAL CELL ASSEMBLIES CAN CONVEY ABSTRACT MEANING

When an animal navigates in two-dimensional environments, neurons activated at crossroads become tied to different routes, emerging as "omnidirectional" cells. In the physical environment, omnidirectional cells serve as junctions to establish maps[105]. Multiple episodes of common junctions can free neurons from their physical context[106] [78]. Neurons that belong to a specific omnidirectional assembly can further define or symbolize the meaning of an item. Conversely, emergent levels of omnidirectional assemblies can become part of higher order representations[107]. These assemblies have a relational origin and do not necessarily vary with the conditions of their creation.

Thus, navigation experiments provide an empirical basis for Deacon's ideas [55]. This line of research also supports the paradigm *information.expectation* transition of Shannon's information to non-Shannon's "aboutness" information. Awareness of meaning can be realized as:

'information.expectation' information.expectation

relationships among higher order representations generated by omnidirectional neuronal assemblies or, in applying this blueprint template to hippocampal activity:

omnidirectional 'afc.ana' afc.ana ≈ realization of meaning[108]

However, as we will see in Predictive Coding, realization of meaning should be updated to:

'expectation.information' expectation.information

and, among higher order representations generated by omnidirectional neural assemblies:

omnidirectional 'ana.afc' ana.afc ≈ realization of meaning[109]

THETA OSCILLATIONS CROSS-COUPLING CAN GENERATE HIGHER-ORDER CELL ASSEMBLIES

There is another way of developing neuronal assemblies with the ability to represent abstract concepts: the time-dependent coordination of neuronal activity. I refer to *"the qualitative essence of neuronal synchrony: coupling through time by some invisible links"* [78].

Navigation studies that correlate location with time provide a window to study the phenomenon. To this end, this very concise explanation failed to include a fundamental aspect of hippocampal activity: theta oscillations[110]. Hippocampal theta oscillations (6-10Hz in the rat) maintain a constant rhythm while the animal performs an activity. The rhythm can correlate with behavioral navigation.

For this review, I will resort again to Buzsáki [78]. Place cells have long and heavy activity tails and, in consequence, the spikes of many neurons overlap in time. As the animal moves, the activity of multiple neurons representing overlapping place fields shifts together in time during successive theta cycles[111].

Theta oscillations of neuronal activity allow the coordination of neural spikes over time. As the animal moves, the spikes of place cells also shift systematically relative to the phase of the ongoing theta oscillation[112]. The temporal coordination of neuronal spikes by theta oscillations fosters spike-timing-dependent plasticity[113] and the development of connections driven by the theta phase, in addition to those related to closer physical locations[114]. Neurons firing in the same phase of theta oscillations can be become more closely coupled, irrespective of the animal physical location. More precisely, phase-related connectedness allows for the emergence of abstract relationships based on time, and not necessarily location (Fig. **4**).

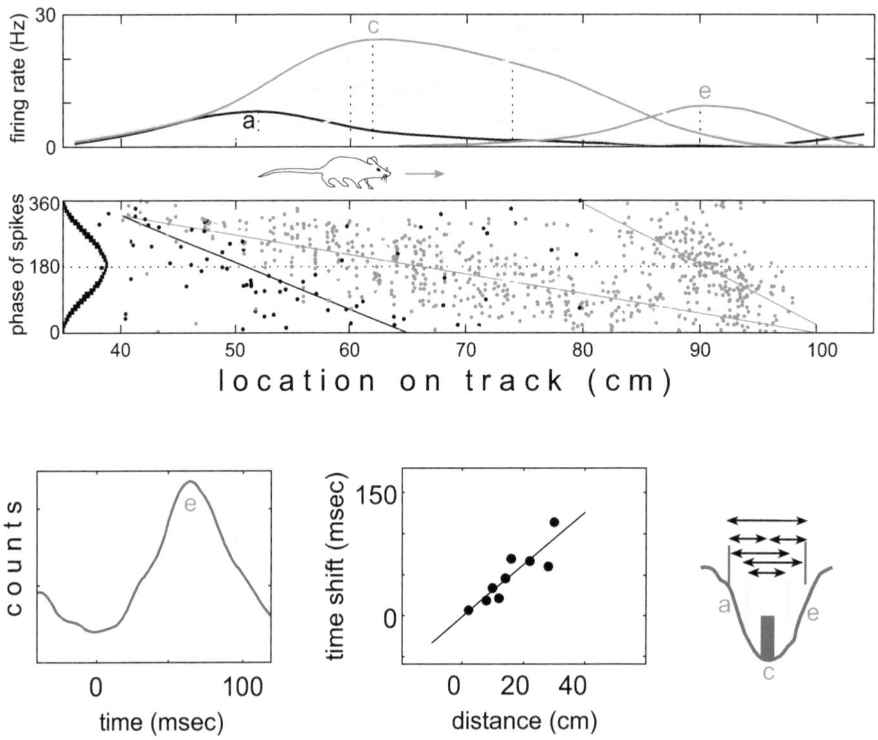

Fig. (4). Representations of spatial distances by temporal correlations between spikes of hippocampal neurons. It shows the firing range of five place neurons (a-e) while the animal travels (axis x) and the phase of the theta cycle (axis y). Highest firing rate happens at theta nadir. Lower left shows that the firing rates of neurons d and e have a temporal relationship. Lower right shows temporal relationships between spikes of different place cells. Reproduced with permission from Elsevier Science [82].

Thus, neuronal plasticity can be determined not only by neurons representing adjacent places in represented maps, but also by higher-order non-adjacent neurons firing at the same phase of several theta cycles[115]. Why is this relevant? It describes how neuronal plasticity could generate cellular assemblies that do not vary with their conditions of origin. I will describe later how human consciousness is thought to depend on long-distance reverberating circuits, including thalamocortical and corticocortical circuits[116].

Relational abstract representation is consistent with theories that explain human memory [83]. Memory retrieval prompts the encoding of new memory traces across hippocampal sites. The overlap of memory traces allows for the decontextualization of memory and the emergence of semantic memory that is independent from its origin[117].

Information may thus be carried not only by spatially distributed cell assemblies, but by a distributed temporal code depending on the reverberation of temporally coinciding signals, functionally decoupled from spatial signal contents [84]. Irrespective of giving to temporal code a specific causal role for the generation of conscious states, contemporary biophysics points towards states of abstract representation, of relational origin, which are consistent with the fundamental ideas of this blueprint.

ACTIONS GENERATE CONCRETE AND ABSTRACT MAPS

This line of experiments has an implicit corollary: motor actions generate maps. In the absence of motor activity, somatosensory information cannot incorporate spatial relationships among different stimuli. Movement-induced sensation lies at the basis of representation in the physical world. No perception exists without action [78]. Experience modifies neuronal connectivity, but experience should be defined as knowledge and skill gained by direct action. The same ideas should apply to abstract map-making by omnidirectional cells and time-based connectivity. It all flows from the animal actions.

Consciousness serves as a platform for adaptive global actions. Morsella and colleagues [18, 28, 29] extensively discuss consciousness from an action-based perspective serving the somatic nervous system. Considering that consciousness only controls the actions of the muscle-skeletal system, these authors approach consciousness from the perspective of the requirements of adaptive action control instead of perceptual analysis[118].

Adaptive behaviors enacted by the muscle-skeletal system require a global view of activity options. A conscious field encompassing all of which the organism consciously perceives at one time allows for the integration of multiple systems. It enables the organism to make choices among possible global actions[119].

These ideas apply to the development of abstract representations. "*Nothing in the brain makes sense, except in light of behavior*" [78]. Again, according to the IIT, even inactive systems can be conscious[120], in contrast with the review of the literature so far presented here. The IIT is mistaken. Consciousness is something we do, and qualia the product of internal changes that occur during the action.

CONCLUSIONS

Navigation studies provide a mechanism for information to be abstracted from spatial and temporal perspectives. Relationships among place cells and place fields are consistent with *information.expectation*[121]. The realization of information by omnidirectional assemblies is consistent with states of abstract representation, of relational origin.

Likewise, the realization of temporally coded information is consistent with *information.expectation*. The abstraction achieved by functional decoupling of spatial signal contents is consistent with thermodynamic decoupling. Actions generate abstract cognitive maps necessary for awareness. All this fits a subject-centric, action-based theory of consciousness.

NOTES

[92] The hippocampus internal recurrent system has the computational properties of a recursive organization and meets the theoretical requirements of an "autoassociator" [78]. An associative network has a memory capacity (the maximum number of patterns that can be stored and retrieved) and a content addressability (the ability to recall a whole episode from a cue of a small part of the original information). As reviewed by Buzsáki [78], the autoassociator and its synaptic matrix represent a strongly connected, directed, and weighted graph. Hippocampal cell assemblies representing the same information are distributed randomly over entire regions. The hippocampus distributes memory over a large space, with only a fraction located at one physical location such as a synapse.

[93] The brain of humans and other vertebrates has two hippocampi, which are seahorse-shaped structures located inside the temporal lobes and known for the consolidation of memory and spatial navigation. While the brain neocortex is modularly organized to extract statistical regularities of sensory information, the hippocampus provides a large storage space with randomly organized connections, which allows building event sequencing through the spatiotemporal context of stored information [78].

[94] The hippocampus computes the allocentric (other than self centered) space, like in landmark or celestial navigation.

[95] Pyramidal cells are the main excitatory neurons in the cerebral cortex, hippocampus and amygdala.

[96] I consider knowledge of location a form of awareness within a spectrum of consciousness. This specific awareness must be integrated and realized in the overall experience.

[97] At this level, the hippocampal mechanism for realizing knowledge of location has a relational origin, given by a relationship between the information presented by centrally active neuron(s) and an underlying field of possibilities defined by an assembly of less activated cells. Note that the relationship is the product of active neurons, a finding contradictory to the prediction of the Integrated Information Theory that inactive systems will convey consciousness.

[98] Head direction neurons are orientation-specific and location-invariant, whereas place cells in the hippocampus are orientation-invariant and location-specific.

[99] The entorhinal cortex is located in the medial temporal lobe. It functions as a hub for memory and spatial navigation and the main interface between the hippocampus and the neocortices. The entorhinal cortex contains a map of the spatial environment in rats. Relocation of the animal to another environment remaps hippocampal patterns of spatial activity. The activity of grid entorhinal cells will change when a complete hippocampal remapping occurs; otherwise they retain their original pattern of activation. The mechanism signaling knowledge of position remains the same: the spike of activity of specific neurons within a context of a specific pattern of less activated hippocampal neurons. The strength between cell connections represents distances in the actual environment. Same cells can be used to represent different environments, but their relationships will vary on map-by-map basis. A large enough change in the environment will cause a drastic activity change in place cells activity. A large portion of cells active in the initial environment becomes silent in the second. Among those that continue active, firing locations become completely reorganized in space [81].

[100] Edward Tolman proposed in 1948 cognitive maps to represent relative spatial locations in the context of a mental frame of reference [303].

[101] Grid cells are organized in modules tuned to oscillations in the local field potential. Theta rhythms are implicated in the generation of the grid pattern (more of this later). Grid cells in geometrically defined modules are modulated by the same theta frequency [81].

[102] In his excellent chapter, "Navigation in real and memory space", Buzsáki provides a concise review of the literature on this theme [78].

[103] The perirhinal cortex (PRH) is located at the medial temporal lobe and bordered by the parahippocampal and entorhinal cortices. It encodes larger portions of the environment, independently of body or head location, within the context of a task. In contrast to the characteristic transient hippocampal activity, PRH neurons provide sustained firing to keep track of the animal location on larger maze segments. PRH neurons phase-lock with hippocampal theta oscillations and the hippocampus seems to exert top down influence on the PRH: *"From a conceptual viewpoint, both maze chunks and other items could be argued to be objects, defined as composites of informational elements bound by multisensory and spatiotemporal properties"*[80].

[104] This is the alluded rabbit hole: building up consciousness from the bottom-up brings up emptiness; pressing top-down loses qualia in nothingness. Is it the chicken or the egg? Like the old childhood riddle, the answer is found in the long view of evolution. There were eggs much before there were chickens. There was inference about the environment, according to the principles of statistical physics, much before human consciousness.

[105] Route-finding optimization is based on omnidirectional explicit place cells and synaptic strengths between them.

[106] Note that these intersections are the reason why realization of information provided by the relationship between a central spike of neuronal activity and an overall activation of a place field goes beyond a gradient of activity but can present itself as a qualitative component. In the context of multiple maps junctions, the firing of a single neuron is related to the context of firing of all other neurons actively representing potential alternatives intersecting across multiple maps. This can be consistent with realization of information across mathematical structures of multiple dimensions [279].

[107] As we will see later, the same applies to the thalamic-cortical unity in the dGW version of the Global Neuronal Workspace (GNW) theory.
[108] *'abstract-field-center.abstract-neuronal-assembly'*
abstract-field-center.abstract-neuronal-assembly
≈ realization of meaning
[109] *'abstract-neuronal-assembly.abstract-field-center'*
abstract-neuronal-assembly.abstract-field-center
≈ realization of meaning
[110] Hippocampal theta rhythm (6-10Hz in rats) is the most prominent rhythm of hippocampal activity; a sustained rhythm associated with explorative navigation and considered the major organizer of the hippocampal-entorhinal cortex.
[111] For all place cells, the highest firing rate occurs at the nadir of the theta cycle.
[112] Phase precession.
[113] A strong supra-threshold synaptic input increases the synaptic strengths of sub-threshold inputs a few tens of milliseconds earlier. The mechanism plays a central role in state dependent learning of cortical networks (more of this in the chapter of time perception).
[114] The temporal relationship between neurons reflects the strength of their synaptic links. As the animal moves forward, a new cell assembly dominates each theta cycle. However, assembly members also contribute to the representation of past and future locations in multiple theta cycles. Therefore, information spans more than a single assembly in one theta cycle. The relationships among several theta cycles allow for the generation of a spatiotemporal context based on the temporal sequences among cycles.
[115] As the animal moves forward in the field, multiple cell assemblies are active at each theta cycle overlying representations of past, present, and future locations so that physical distances are also reflected in the temporal sequences in the cycles. This mechanism provides a spatiotemporal context, necessary for episodic memory. Not only temporally adjacent items within the theta cycle, but also non-temporally contiguous items matching positions in the ups and downs of several theta-cycles can be reinforced by synaptic plasticity.
[116] Every part of the neuron can function as a resonator-oscillator. Neurons have a tremendous computational power. They are complicated resonators, capable of response over a wide range of frequencies. Information *"can be embedded in the dynamic change of temporal synchrony even without an alteration in the discharge frequency ...information stored in a given phase of the oscillatory output of one structure could be effectively read out by the target structure, provided that the faster oscillation in the target is phase-locked to the input and the combination of the two oscillators sufficiently depolarize the neuron" [78].*
[117] And also support the recontextualization of memory over time through the substitution of the original memory context with memory retrieval. Recontextualization can be seen as changing in neuronal assembly "expectations" providing the context for a specific memory file.
[118] An approach consistent with Noë [48] view of consciousness as action.
[119] In Morsella's words, conscious contents are "*well-crafted representations occurring at a stage of processing between sensory analysis and motor programming.*" An action-

based view of consciousness implies that phenomenal experiences are the drivers of goal directed actions. It suggests an adaptive advantage of phenomenal consciousness as early as the evolutionary onset of global goal directed actions.

[120] According to the IIT: *"inactive (i.e.* silent) elements generate information and thus contribute to specifying the shape of the quale. Along the same lines, a somewhat counterintuitive prediction stemming from the IIT is that if elements within a complex are inactivated, rather than merely being inactive, experience should change, although the firing pattern is the same" [77]. The same firing pattern will represent different information depending on the repertoire of potential firing, but the meaning of "potential" in inactive *vs*. inactivated is blurred in the IIT.

[121] "*Explicit absence of neuronal spikes (anticorrelation) is as important as the spiking activity of assembly members*" [78].

CHAPTER 11

Awareness of Time

Abstract: Humans sense the passage of time as a subtle form of experience defined by an absence. Studies of interval timing often focus on the individual's assessment of a period with no stimulation between two auditory stimuli. Time is an inherent part of Shannon's information. Brain function contains an invariant time-dimension. Every cortical circuit seems to have an inherent computational ability for timing regulated by time-dependent changes in synaptic and cellular properties. Local networks affect qualia of time experienced both in visual and auditory experiments. State-dependent local learning and oscillatory phase shifts support a role for early cortical processing in time discrimination. Brain oscillatory phases predict conscious perception. The temporal cortex adjusts its own oscillatory phase, mapping its window of analysis of incoming time-sensitive events. These phase adjustments support the existence of active brain-centric expectations in time perception and other conscious experiences.

Keywords: Alpha power, Auditory Consciousness, Brain-Centric Expectations, Early Cortical Processing of Information, Entrained Neuronal Stimulation, Interval Timing, Multimodal Integration, Near-Threshold Perception, Neural Assemblies, Oscillatory Phase, Pre-Stimulus, Scalar Timing Theory, Shannon's Information, Short-Term Synaptic Plasticity, Striatal Beat-Frequency Model, State-Dependent Expectations, Temporal Dynamics, Time Perception, Visual Cortex Adaptation, Window of Analysis Calibration.

Authors often fail to recognize qualia as the subtle forms of experience associated with the passage of time. I became interested in time perception for this exact reason: this experience does not carry the overt qualities traditionally related to qualia. Instead, it can be defined by an absence. For example, studies of interval timing[122] often focus on a subject's ability to assess a period with no stimulation between two auditory stimuli.

The experience of time does not directly emerge as a product of external stimulation, but due to internal mechanisms. Experiments must separate the direct subjective product of external stimulation, such as hearing a noise or seeing a flash of light, from the experience of time itself, which emerges from internal mechanisms employed to process the external stimulation.

Paulo J. Negro

A BRIEF REVIEW OF THE PHYSIOLOGY

Organisms must quantify time to survive[123] [85, 86]. As time is an inherent part of Shannon's information, information processing in the brain is implicitly coupled with the *"invariant time-dimension"*. Temporal information guides behavior in different time scales. No sensory organs or primary cortex exist with the specific purpose of time perception, but organisms extract temporal information from all sensory modalities[124] (Table **1**).

Table 1. Different time scales of adaptive behavior [85].

Scale	Actions
Microseconds: neuroprocessing	Binaural hearing (differences in action potentials originated from the left and right ears) Echolocation (creation of neural images of the environment)
Milliseconds: interval timing	Object interception or collision avoidance Speech perception and articulation Execution and appreciation of music and dance
Seconds to minutes: interval timing	Foraging, decision-making, sequential motor performance Multiple-step arithmetic, associative conditioning (Highly influenced by attention and memory)
Circadian rhythms: 24h oscillations	Optimizing biological activity during the day

Several brain regions participate in the multimodal integration of information duration, such as the posterior parietal cortex, the superior temporal sulcus and the dorsal medial temporal area. Time perception requires multimodal integration[125] [85], which includes a hippocampal role for the experience of the living present and fringe consciousness[126] [83]. Interval timing supports predictions of the onsets and offsets of environmental events. Most interval timing models assume a combination of clock, memory, and decision-making.

AN INTRODUCTION TO INTERVAL TIMING

According to the Scalar Timing Theory (STT), memory forms the basis of a scalar property[127] through the transmission of information from an accumulator to a memory store[128] [87, 88]. The Scalar Timing Theory evolved into the Striatal Beat-Frequency (SBF) model [88]. The SBF is a biologically grounded framework for the STT [89].

Very concisely, cortical neurons or neuronal assemblies oscillate at a stable, but different, frequency over time. They are most likely located in the prefrontal cortex (PFC), and possibly in other regions such as the supplementary motor area (SMA), the posterior parietal cortex (PPC), and sensory cortices. Cortical oscillators are thought to phase-reset at the beginning of the time to be timed, and neurons located in the striatum[129] read their status[130] [88, 89].

Current interpretations of the literature support the idea that specific cortical oscillators may not be necessary because any stable pattern of neural activation might support interval timing[131] as long as there can be reliable coincidence detection[132] [89].

LOCAL NETWORKS CAN ENCODE LOCAL TIME

Johnston and colleagues investigated the effect of localized temporal mechanisms in the visual cortex on the perceived duration of a visual stimulus [90]. The authors took advantage of the anatomical separation of the visual system fibers[133] that start at the retina and is projected to the visual cortex.

They compared two visual stimuli with the same duration[134] after inducing the local adaptation to one of them. They pre-exposed a specific location of the visual field to a "standard" stimulus and compared its perception to a non-pre-exposed location. Each location was connected to a different region of the primary visual cortex. This clever study design identified the effects of local neural adaptation on the perception of time intervals[135].

Their subjects perceived the stimulus to have a shorter duration in the pre-exposed location. Most interestingly, despite the subjective experience of different durations they correctly matched the stimulation onset and offset in both locations. Changes in apparent duration after adaptation did not result from the altered latency at onset relative to offset[136].

Thus, complex recurrent networks may play a role in the flexibility, diversity and complexity of sensory timing. As identification of the onset and offset of the visual stimuli did not vary, these results suggest an effect on the actual experience of time itself. If we accept time perception as a subtle form of qualia, then, the adaptation of local networks affects qualia of time experienced[137].

SHORT-TERM PLASTICITY CAN EXPLAIN LOCAL EFFECTS IN INTERVAL TIMING

Every cortical circuit seems to have an inherent computational ability for timing, which can be realized locally [91]. According to Buonomano, short-term synaptic plasticity[138] triggered in the early auditory cortex during simple auditory discrimination tasks can create time-dependent changes in the network and encode spatiotemporal information [92].

He compared two intervals presented 250ms apart. When tones were presented with different frequencies, interval discrimination was not affected. However, when tones were presented with the same frequency, the ability to discriminate the longer tone was impaired. Similar tones activate the same local neural network[139].

Similar frequencies are supposed to impair interval discrimination by interfering with the state of the network and by not allowing the local circuitry to reset for the next tone. Different frequencies cause less interference because low frequency tones do not strongly change the state of the high frequency local network.

Buonomano's findings match those reported by Johnston's[140]. Both authors provide evidence for a role for early cortical processing in time discrimination[141]. Cortical networks time-dependent changes in synaptic and cellular properties can influence their population response to events in a history-dependent manner.

OSCILLATORY PHASE ADAPTATION FRAMES BRAIN-CENTRIC TIME EXPERIENCES

The brain changes its time processing pattern to actively compensate for temporal delays [93]. I describe next the work of Kösem, who directly addressed the issue.

Kösem assessed whether the phase of an entrained[142] neural oscillation directly informs the conscious experience of time.[143]. The researcher phase-locked a 1Hz oscillatory activity to the visual and auditory cortices of her study subjects[144]. However, the phase of the 1Hz oscillatory component did not remain stationary during a lag-adaptation experiment.

Neural responses were evoked by further stimulation arriving at different phases of the 1Hz oscillation. When auditory stimulation preceded vision stimulation by 200ms (A200V) and, in the opposite case, when vision stimulation preceded auditory stimulation by 200ms (V200A), phases of the 1Hz oscillation shifted in opposite directions in the visual and auditory cortices[145]. No phase shift was observed in the control condition of simultaneous audio-visual stimuli [93].

Auditory and visual evoked potentials did not explain changes in perceived timing[146]. Phase shifts of the auditory responses predicted shifts of subjective AV simultaneity[147]. There was no systematic relationship between visual responses and subjective AV timing. In summary [93]:

> *"The auditory cortex temporally calibrates its window of analysis with respect to vision and that event timing linearly maps on the phase of entrained neural oscillations"*

In other words, the subjective perception of simultaneity was generated by adaptive changes of the temporal cortex. The perception of simultaneity was brought by adjustments in the phase of temporal cortex oscillations. In the language of this blueprint, adjustments of oscillatory phase constitute active "expectations" of the temporal cortex [93]:

> *"Our results suggest that, in cortex, the phase of neural oscillation may provide an automatic means to flag events in a brain's referential time – i.e.* provide the needed brain-centric view of time".

BRAIN-CENTRIC EXPECTATIONS PREDICT CONSCIOUS PERCEPTION

Pre-stimulus oscillatory activity predicts conscious perception in low contrast or near-threshold stimulation studies. The phase of brain activity in relevant areas directly correlates to how well it processes information inflow [94]. These phase effects closely align with stimulus onset or occurs slightly before stimuli. Phase modulations are observed for expectation [96] and attention [97] when the timing of the stimuli is predictable.

The relationship between spontaneous EEG alpha phase and the amplitude of event related potential[148] has been reported for decades [98]. Top-down alpha-band oscillations seem to control the disengagement of task-irrelevant areas [99]. However, although the power of neural oscillations[149] affects visual perception, the effects of the pre-stimulus phase bear a closer association to the actual coding and processing of visual information. As reported by Busch and colleagues [98], the oscillatory phase accounted for at least 16% of the detection performance and allowed for the prediction of performances on a single-trial level. In contrast, the effect of the alpha power accounted for 12% of performance variability. Power and phase had seemingly independent effects.

CONCLUSIONS

Active shifts in brain oscillations correlate with the perception of interval timing [93 - 95]. The subtle experience of time was not related to evoked response fields, but to adjustments in the oscillatory activity of neuronal assemblies.

Cortical circuits learn timing. Recurrent network connectivity enables state dependent hidden expectations[150]. Local realization of information in time perception is consistent with the concept of localized consciousness-information.

State dependent local learning and oscillatory phase shifts support the importance of active expectations in time perception and other conscious experiences. These observations can be considered consistent with Bayesian mechanisms for conscious perception already present in early components of the brain decision network.

NOTES

[122] Interval timing is "*the ability to perceive, store, encode, and reproduce temporal intervals ranging from few 100 milliseconds to minutes*" [86]. In computation: a digital circuit used to determine the time interval between an initial trigger pulse and subsequent logic states (a true or false binary signal) that appear after a predetermined delay.

[123] "*An accurate representation of time-dimension in the neuronal circuits is required for a successful interaction of the brain with the four-dimensional physical world*" [86].

[124] In common among sensory modalities: sensory transduction of physical information into action potentials, projection to the primary sensory regions of the cerebral cortex, processing of information by cortical and subcortical circuits involved by the sensory input, higher order processing for perception, learning, memory, and voluntary motor action. Of note, the auditory system has higher capacity for time quantification [85].

[125] It can be difficult to differentiate cross modal bias of unisensory awareness from genuine multisensory properties (multisensory objects and events, and multisensory scene perception) [304].

[126] Hippocampal damage can cause distortion of time perception for intervals longer than 20 seconds. Gross distortions can be prominent for longer intervals. Fringe consciousness is the background awareness for the context of experience [80, 81].

[127] The Scalar Timing Theory is an information-processing version of the Scalar Expectancy Theory. A time-keeping system must be able to target the duration to be timed (the mean estimated intervals must be close to real duration) and the variability (dispersion) of the series of trials must be as low as possible. The variability is represented in the Weber fraction $E_T=kT^N$, where E_T is estimated time, T is the physical time, *K* a constant for the intercept, and N expected to be 1 in the Scalar Expectancy Theory. The theory stipulates that the proportion between variability and mean is scalar, *i.e.*, it is supposed to be constant when the psychometric is obtained with different target durations, in a time invariance principle [88].

[128] The original theory postulated an internal clock consisting of a pacemaker sending pulses, stored in an accumulator. A starting signal would close a switch in the beginning of the time to be timed, making pulses reaching the accumulator to be integrated as a function of time. The pulse count "*could serve as a subjective estimate of time*" [89]. The relationship between pacemaker and accumulator provides a measure for interval timing. The theory has one central hypothesis: the existence of a single internal clock, a pacemaker-counting device.

[129] The striatum is a subcortical portion of the brain. In primates, the dorsal striatum includes the caudate and the putamen, and the ventral striatum includes the nucleus accumbens and the olfactory tubercle. Striatal medium spiny neurons (MSNs) act as the detectors, firing with the identification of specific oscillatory patterns.

[130] The arguments for SBF: 1. The striatum is the main input system for the basal ganglia; 2. Striatal neurons receive input from up to thirty thousand cortical and thalamic units; 3. The basal ganglia is considered a filtering system and MSN neurons need a large

input of coherent stimulation to fire; 4. About 20% of dorsal striatum cells firing depend of temporal modulation (*e.g.*, some assemblies firing 10 seconds after a stimulus, others 40 seconds after a stimulus) [88].

[131] Possibly, working memory and interval timing "*can originate from the same oscillatory processes such as gamma and theta oscillations, and phase-amplitude coupling between these frequency bands*" [89].

[132] Most likely, a hybrid model with a partially distributed timing mechanism integrated by core structures – such as the cortico-thalamic-basal ganglia circuit – best explains the temporal performance variability in different contexts. Motor timing, for example, seems to depend on the activity of recurrent cortical networks that display a self-sustaining activity. From an action-based perspective, the medial premotor area seems to contain different populations of duration-tuned cells. These cells become activated depending on duration of intervals. These cells can represent the duration of events, placing the medial premotor area as part of a core timing circuit. Neurophysiological recordings show that the medial premotor cortex (MPC) uses multiple encoding strategies to represent different aspects of temporal structures [305,308]. Therefore, the MPC takes part in a core-timing network. This network uses interval tuning as a signal to represent temporal processing in behavioral contexts involving the explicit quantification of time.

[133] Visual temporal mechanisms are differentiated early in the retina, have small receptive fields (less than half a degree of the visual angle for the majority of foveal V1 receptive fields) and project to spatially localized regions of the primary visual cortex.

[134] Exposure to a sinusoidal grating was designed to induce adaptation on one side of the central visual fixation point. More precisely, the duration of one of the standard stimulus, shown on the adapted side, was fixed (600ms), while the duration of the comparison, shown on the unadapted side, was varied between ±1 octave of the standard duration (300–1200ms). Both stimuli were Gaussian windowed sinusoidal gratings. In any run of trials, the orientations of the comparison and standard gratings were randomly set to vertical or horizontal. Each point on the psychometric function was based on 10 trials and each data point was the average of 4 trial runs. The 50% point on the psychometric function provided a measure of the perceived duration of the comparison required to match the duration of the standard after adaptation. In the temporal order experiments, the subjects compared the onset (or offset) of the stimuli against the onset of a 30ms square wave white noise burst delivered binaurally. The subjects reported which occurred first. In the interval generation experiment either the standard or the comparison grating was shown in isolation (for 600ms). Finally, subjects compared the duration or temporal frequency of a flickering Gaussian comparison against a standard. Subjects reported which of two intervals contained the longer stimulus. Observers attempted to generate an interval that matched the stimulus duration by pressing a lever [90].

[135] The perceptual latency of the onset and offset of a standard pattern with or without adaptation.

[136] To dissociate the effects on duration perception from any manipulation of the time course of sensory processing.

[137] "*...the recruitment of areas for which time perception appears not to be the primary function may suggest a more distributed form of neural organization for duration perception. The number of brain regions implicated in timing judgments is indicative of*

the complexity of the duration judgment task. Some of these areas might be involved in selecting and marking the starting and finishing points for interval judgments and may not directly encode event duration. Both the stimulus modality and the nature of the timing task may be critical to which brain areas are recruited. Adaptation might influence the timing mechanism at a number of levels. The spatial specificity of the temporal frequency adaptation effect described here, however, suggests that we are targeting an early component of the decision network" [90].

[138] The *"hidden activity state"* in simple discrimination task [92].

[139] The auditory system is tonotropically organized.

[140] Keep in mind that recurrent connectivity can change local circuitry "expectations" (more of this in interoception and predictive coding).

[141] *"The main input layer of the sensory cortex is layer 4, but neurons in all layers can be activated by the tone, and there is a high degree of both feedforward and recurrent connectivity within any cortical circuit... because of short-term synaptic plasticity (the hidden state) the strength of many synapses may be different at the arrival of the first and second tone resulting in the activation of distinct subset of neurons... The difference in these populations can be used to code for the interval between the tones. State-dependent network models predict that as information flows through different cortical areas, the encoding of temporal and spatiotemporal information may increase, but could begin at early sensory areas, such as the primary auditory cortex."*

[142] Brain oscillations are modulated by sensory inputs by phase reset and entrainment. In neural entrainment, sensory cortices follow environmental rhythms. Entrainment allows the alignment of the cortical processing with external events. Normal volunteers develop a robust 1Hz auditory and visual cortical frequency-tagging response to rhythmic desynchronized audiovisual (AV) stimulation. In this study, this was recorded by magnetoencephalography (MEG) and co-registered with a structural magnetic resonance imaging (MRI). The author submitted those volunteers to experimental blocks of 65 AV streams of stimuli at an average rate of 1Hz and tested for a series of simultaneous (S) or desynchronized AV events. Visual stimuli were projected at a 60Hz refresh rate, while auditory stimuli consisted of 16ms duration white noise. The analysis was restricted to regions of interest at the right auditory and visual cortices.

[143] A200V (audition leading vision by 200ms) or V200A (vision leading audition by 200ms). The stimulus onset asynchrony between successive auditory or visual stimuli was randomly selected from a normal distribution with a mean of 1s and a standard deviation of 100ms. Each stimulus onset asynchrony had 95% chance to fall between 804ms and 1196ms. After each presentation of an AV pair, participants were asked to judge which of the sound or visual stimuli appeared first.

[144] Event-related fields (ERF) were calculated by averaging 15 trials at the beginning and at the end of a lag-adaptation experimental block. The stimulus onset was locked to the sound onset for auditory ERF, and to the visual onset for visual ERF. Event related field (ERF) is a measure of the brain direct response to a stimulus. In order to see the brain response to a single stimulus it is necessary to repeat the stimulation many times and average the results together, eliminating noise and generating a response in the form of a wave. In this study, the authors focused on the M100, which is a large negative wave peaking between 80-120ms after the onset of a stimulus. It is involved in perception and

its amplitude is strongly dependent on time and characteristics of the stimulus.

[145] During lag-adaptation auditory analysis may be delayed (a negative shift of the phase of auditory entrained oscillation, *i.e.* the 1Hz response in auditory cortex shifted forward in time) and the visual analysis advanced in time (a positive shift of the phase of visual entrained oscillation, *i.e.* the 1Hz entrained response in visual cortex shifted backward in time). Lag-adaptation to simultaneous AV stimuli (S) did not affect the phase of the entrained 1Hz oscillation in neither cortices.

[146] Evoked responses did not impact the neural oscillatory phase shifts, which the researchers considered active brain adaptations. Overall, no significant differences in the auditory and visual evoked responses profiles were observed in all lag-adaptation experimental conditions. Visual evoked responses were not modulated by auditory stimuli. However, auditory evoked responses were modulated by the presence of visual stimuli. Modulations of the auditory evoked responses were steady throughout the experiment and did not differ between the beginning and the end of a given experimental block. Because changes in perceived timing were related to changes in neural timing of auditory and visual cortices during lag-adaptation, the latency of the evoked responses did not appear to capture these changes.

[147] Point of subjective simultaneity.

[148] Evoked related potentials are obtained with EEG (electroencephalography), in contrast to evoked related fields, which are obtained with MEG (magneto-encephalography).

[149] While power modulation is more representative of sustained excitation of target regions, phase likely imply increased excitability at certain points in time. Low alpha band power predicts perceived in comparison to non-perceived near threshold stimulus. Alpha oscillations likely modulate the excitatory state of cortical regions, top-down suppressing task-irrelevant areas [94].

[150] Recurrent neural networks with self-perpetuating activity have chaotic dynamics, which are sensitive to noise and not easy to model. These networks express dynamic attractors [95]. However, instead of converging to a fixed point, neural trajectories are best seen as representing spatiotemporal objects.

CHAPTER 12

Interoception

Abstract: Interoception consists of sensing one's own physiologic conditions, including heartbeats, gastric filling, and bladder fullness. The experience emerges from neuronal activity at the insula. The brain represents primary interoception in the posterior insula and integrates it with emotional salience in a posterior to anterior insula activation pattern. The posterior to anterior instantiation of interoceptive information provides evidence that consciousness indeed represents the product of abstractions across the brain hierarchical levels. Different abstraction patterns refer to distinct levels of representation of tissue-specific and whole-body homeostatic states, from basic interoceptive experiences to a more abstract representation of wellbeing, integration of exteroception, emotional awareness, and cognitive processing. The brain successively realizes these qualitative differences along the axis that links the posterior to the anterior insula. Interoception and its derivative emotional states fit the concept of realized information in the context of potential states represented across system hierarchies. These studies and considerations validate the paradigm of embodied abstractions. Research in the field of interoception often supports a role for strange loops in the realization of information. Interoceptive awareness arises from the fusion of subject and body, where the body plays the role of a building block of the self it represents. This pattern is consistent with consciousness as action, and with Damasio's views of consciousness as the product of a reaction of the organism to an object as the organism regulates itself to process this object.

Keywords: Anterior Cingulate Cortex, Awareness of Heart Beats, Bladder Fullness, Computational Hierarchy, Damasio, Embodied Abstractions, Emotional Awareness, Experience of Body Ownership, Feeling of Agency, Homeostatic Salience Integration, Interoception, Insula Hierarchical Organization, Insular Cortex, Insula Instantiation, Meta-Representations, Multisensory Integration, Primordial Strange Loop, Rubber Hand Illusion, Realization Across Levels of Abstraction, Sense of Self.

I left out of the time perception review an important structure: the insula. As reviewed by Craig [100], timekeeping studies point towards a role for the joint activation of the anterior insular cortex and the anterior cingulate cortex in time perception. The junction between the anterior insular cortex and the inferior parietal cortex also plays a role in time perception. In fact, the anterior insular

Paulo J. Negro

cortex is associated with the *"rhythmic temporal progression of laden moments"* that characterizes human appreciation of music. It possibly participates in the subjective time dilation of strong emotional experiences. My omission had a purpose: the insula takes part in a wide variety of experiences of awareness, which includes the awareness of time.

A BRIEF REVIEW OF THE ANATOMY AND PHYSIOLOGY OF THE INSULA

Interoception is the sense of the physiologic condition of the body, the awareness of heartbeats, gastric filling, and bladder fullness[151]. It localizes to a specific region of the human brain, the insula[152]. More specifically, the posterior third portion of the dorsal insula is the primary cortex for interoception in humans [101].

The insula plays a critical role in the integration of interoception, focal exteroceptive attention and cognitive control. It matches information from the internal milieu[153] with information from other brain regions about intended changes and expected states on the body.

The insula has a hierarchical organization. Primary interoception is represented in the dorsal posterior insular cortex, while the mid-insula and the anterior insular cortex incorporate salience in a posterior to anterior pattern of integration of interoception with emotional awareness [100 - 104].

For example, functional studies show the preferential activation of the posterior insula in heartbeat awareness and sensation of fullness [100]. Interoception involves the perception and the elaboration of largely afferent vagal[154] information about the state of the body. The posterior and middle aspects of the insula are strongly connected with the somatosensory cortex. They receive nociceptive, thermal and visceral information [103].

In addition to interoceptive stimulation, proprioceptive[155], motoric and hedonic stimulation activate the human mid-insula [101]. Activation occurs not only during movement, but also by the perception of movement in the rubber-hand illusion[156]. These findings lend support to the interpretation that middle regions play a role in the feeling of agency and awareness of body control [100]. This pattern of connectivity and activation suggests that the production of subjective feelings emerges from homeostatic[157] salience integration.

The human brain activates the anterior insula during subjective evaluations of pain, empathic feelings, and other states of emotional awareness. The anterior insular cortex has strong connection with the prefrontal cortical and limbic regions, including the anterior cingulate, orbitofrontal cortex, amygdala and ventral striatum [103]. While it can be active during interoceptive awareness, the anterior insula is particularly activated by emotional stimuli, focal attention and detection of perceptual salience[158]. The insula participates in visual and auditory awareness of the moment, perceptual decision-making and in the evaluation of risk, of experience, anticipation and uncertainty [100].

The strong connectivity between the anterior insula and the anterior cingulate cortex[159] fits the idea of the insula as a limbic[160] sensory cortex, and the cingulate as limbic motor cortex. However, the anterior insular cortex has an association with all feelings, and not only those related to the body. It participates in cognitive awareness. The typical co-activation of the anterior insular cortex and the anterior cingulate cortex by emotional feelings may help engage feeling and motivation in a variety of behaviors[161] [100].

THE EMBODIMENT OF ONE'S SENSE OF SELF

As seen above, the anterior insular cortex plays a key role in subjective feelings, as well as in the awareness of sensation and movements. Information across multiple hierarchically organized insula regions generates "meta-representations" of the body homeostatic state [103], which form the template for "feelings" in close connection with the anterior cingulate cortex. The latter is thought to play the part of the volitional agent that modulates those feelings during "global emotional moments" [100].

Human consciousness may have evolved hierarchically, starting with emotions[162], feelings[163], and eventually higher-order cognitive maps. In his influential book, Antonio Damasio describes emotions as embodied patterns of response generated by subcortical neuronal ensembles beginning at the brain stem[164] and automatically engaged without conscious deliberation [105].

Feelings are a second order, but still non-verbal, neural activity. They represent the organism's own changing states as it interacts with, and represents, something else. *"Primordial feelings"* are those providing a direct experience of the organism's living body[165], such as those elicited during the activation of the posterior and mid-insula. Interoception refers to this basic form of consciousness, which forms the basis for the embodiment of one's sense of self[166]. Emotions can be considered actions, and feelings perceptions [105].

THE REPRESENTED BODY AND THE REPRESENTED SELF

Individuals who become aware of their heart rate after infusions of isoproterenol[167] display an activation of the mid-posterior insula in fMRI [106].

In contrast, other evidence indicates that the primary visual cortex is not a direct component of the "content-specific" neural correlates of consciousness of visual experiences. Why the difference?

Note that interoceptive awareness arises from merging subject and body. The represented body is a building block of the self it represents. Interoception is a simpler form of representation[168]. It requires the assembly of a simpler subject and a closer connection between intention[169] and perception. Representations of higher complexity demand more levels of reconstruction and larger neuronal assemblies[170]. All this considered, it comes to no surprise that the primary cortex representing interoceptive experiences is more likely to be included as a direct component of the content-specific neural correlates of consciousness.

Consistently with the idea of consciousness as action, Damasio sees it as the product of a reaction of the organism to an object as the organism regulates itself to satisfactorily process the object[171]. He traces mechanisms of consciousness to single-celled organisms emerging early out of different levels of life regulation[172]. He sees nothing special in neurons that generate consciousness, except for their ability to produce electrochemical signals capable of changing the state of other cells[173], such as other neurons, endocrine cells, and muscle fiber cells.

Damasio emphasize the evolutionary value of mechanisms of consciousness early in the natural history of the planet. Consciousness is seen as a form of embodied cognition as neurons are *"about the body*[174]*"* [107]. Finally, Damasio proposes a mechanism for consciousness in the realization of informational maps[175]. These ideas agree with the concept of consciousness-information and its realization across abstraction levels.

A HIERARCHY OF ABSTRACTIONS IN THE INSULA

More importantly, the insular posterior to mid- to anterior pattern of integration of interoceptive information is consistent with consciousness as the product of hierarchical abstractions across the brain hierarchical organization. The insula [108]:

> *"...instantiates interoceptive representations available to conscious awareness and is associated to processes involving visceral representation, interoception, and emotional awareness relevant to selfhood".*

These different patterns of activation refer to levels of representation of the body homeostatic state, from basic interoceptive experiences to a more abstract representation of wellbeing, integration of exteroception, emotional awareness[176] and associate cognitive processing.

As information moves along from the posterior to the anterior insula, qualitative differences are successively realized. The lamina 1 spinothalamocortical pathway feeds primary homeostatic information to the posterior insula. This information pertains to a multitude of body systems, reaching down to the cellular level. It is experienced as subjective after integration and qualitative realization in the posterior insula.

From the perspective of this blueprint, the experience of increased heart beating, an interoceptive perception, is realized as a qualitatively different perceptual state of emotional awareness through strange loops. However, these strange loops follow a top-down model, rather than bottom-up.

Interoception and its derivative emotional awareness states fit well the concept of realized information in the context of potential states represented across the system hierarchies. The role of the insula in the instantiation of the sense of self in a functional structure of multiple hierarchies of representations fits the paradigm of consciousness as an embodied abstraction.

CONCLUSIONS

Interoception is a basic form of homeostatic and body awareness that merges the represented body and a simple[177] self by making the represented body a building block of the self it represents in what becomes a primordial strange loop.

The hierarchical architecture of the insula is consistent with instantiation of successively higher re-representations. These representations are realized as qualitative perceptual states of awareness: from primary interoceptive awareness to emotional salience to abstract emotional awareness. Such structure is consistent with consciousness as an abstraction emerging from multiple strange loops of hierarchical processual realization.

NOTES

[151] Most studies have gravitated to the perception of heartbeats, which seem to correlate with other interoceptive experiences. The ability to perceive heartbeats at rest indicates a general sensitivity to internal bodily processes [104].

[152] The insula or insular cortex is a portion of the cerebral cortex folded deep inside the fissure separating the temporal from the parietal and frontal lobes. It is disproportionate enlarged in humans in comparison to the macaque monkey [100].

[153] Internal Milieu: a term created by Claude Bernard in 1854, which translates as the "environment within".

[154] The vagus nerve enters the cranium as the 10th cranial nerve, supplying parasympathetic control to most visceral organs (mostly the right vagus nerve) and receiving afferent visceral information from most of the visceral organs (mostly the left vagus nerve). Its actions include controlling of the heart rate, blood pressure,

gastrointestinal activity, mouth movements, gag reflex, and sweating. The vagus provides information to the brain about multiple visceral regions.

[155] Proprioception: information about the position and the movement of the body.

[156] Rubber hand illusion: repetitive touching of an artificial hand and a person's real hand at the same time while keeping the attention focused in the artificial hand leads the person to experience the artificial hand as part of his or her body.

[157] Homeostasis: the regulation and stability of the organism internal conditions.

[158] In Simmons and colleagues [103] study, interoceptive-selective regions were located in an area of the dorsal mid-insula characterized by granular cortex and another region ventral and anterior to the dorsal insula cluster, in the dysgranular insula. Spontaneous resting-state BOLD fluctuations in the dorsal mid-insula interoceptive cluster correlated more with activity in sensorimotor cortex. The activity was considered related to the representation of visceral somatosensation. Spontaneous resting-state BOLD fluctuations in the ventral mid-insular interoception region were significantly more correlated with activity in the anterior insula (adjacent to regions responding selectively in exteroception and emotion conditions) and in the subgenual anterior cingulate gyrus and the temporoparietal cortex, which was interpreted as integrating interoceptive signals with the anterior insula activity for the processing of salience, focal attention and emotional modulation of autonomic activity.

[159] Cingulate cortex: part of the limbic cortex, which includes the cingulate gyrus and sulcus. It is located deep in the medial region of the cerebral cortex, above the corpus callosum.

[160] The limbic system is composed by a heterogeneous group of structures under the cerebral cortex, which are involved in emotions, motivation, long-term memory, and olfaction. The term "limbic" reflects a historical perspective about the brain that addressed cortical structures surrounding the cerebral cortex and the brainstem as a border or limbus.

[161] The left anterior insular cortex is activated mainly by positive and affiliative emotional feelings. The right anterior insular cortex is activated by stimuli arousing to the body [100]. This is interpreted as having an evolutionary origin in the innervation of the heart and the contrast between parasympathetic and sympathetic systems.

[162] Emotions: complex, largely automated programs of actions that are complemented by cognitive programs, but are largely carried out in the body, from facial expression, postures, and changes in viscera and internal milieu [105].

[163] Feelings (of emotions): composition of perceptions of what happens in the body and mind during emotions, such as images of actions instead of actions themselves. In simple organisms capable of behavior, but not of a mind, emotions may be present, but not states of emotional feeling [105].

[164] Brain stem: the lowest part of the mammalian brain, a central trunk that continues to form the spinal cord.

[165] From Damasio [105]: *"Primordial feelings"* are considered the product of a *"proto-self"*. A proto-self refers to the first images generated by the brain and its immediate manifestation of sentience. A *"core self"* is about action, *i.e.* images of an object engaging and modifying the proto-self. While the proto-self and the core self constitute the *"material self"*, the *"autobiographical self"* is considered the social and spiritual self.

[166] There is no consciousness without a self, *i.e.* there is no cryptic intelligent design in Damasio's ideas.
[167] Isoproterenol is a beta-adrenergic agonist (it stimulates beta1/beta2 receptors) and increases the frequency of heart beating. Isoproterenol does not cross the blood-brain barrier and does not directly affects brain function. About 30% of normal volunteers report perceiving their heartbeats. The percentage increases in function of isoproterenol dose and heart response.
[168] Compare the limited details and depth of perceiving heartbeats or bladder fullness with the complexity of visual perception.
[169] Intention: feedback projections in Predictive Coding.
[170] Particularly true for representations requiring multisensory assemblies.
[171] Damasio reports the *"feeling of knowing"* as the feeling of what happens when the organism is engaged with an object. *"An organism can be conscious when the organism representation devices exhibit a specific kind of knowledge: the knowledge that the organism own state has been changed by an object, and when such knowledge occurs with the salient representation of the object."*
[172] *"I believe that minds and consciousness, when they eventually appear in evolution, were first and foremost about life and the life urge within a boundary... As simple as they were and are, single cells had what appeared to be a decisive, unshakable determination to stay alive as long as the genes inside their microscopic nucleus commanded them to do so.... I know it is difficult to imagine that the notions of 'desire' and 'will' are applicable to a single lonely cell... But there they are, by whatever name you may wish to call those features of the cell's behavior."*
[173] *"Neurons did not invent electrical signals... and single-celled organisms can use them to manage their own behavior."*
[174] *"Neurons are about the body, and this 'aboutness' is the reason [Damasio believes] that the genetically engendered will to live of the cells of our body can be translated into the human mind and experiences."*
[175] To Damasio [105,107], the realization occurs through the reciprocal connections between map-making parts of the brain in a "dispositional" space for reconstruction. "Dispositions" are considered abstract records of potentialities or expectations, in the language of this blueprint.
[176] Gur and colleagues [102] meta-analysis of the neural substrates of empathy in humans showed that the activation of the right anterior insular cortex was absent in empathy for positive emotions and positive in empathy for pain and negative emotions, which was consistent with the hypothesis that the right AIC encodes "energy-consuming" negative feelings.
[177] In comparison with other representations and subject assembly that require more levels of reconstruction and larger neuronal assemblies.

CHAPTER 13

Predictive Coding

Abstract: Perception is not a direct image of sensorial input, but a hypothesis explaining sensorial information. The brain hallucinates reality. The ability to infer the causes of sensations requires an internal generative model, known as Predictive Coding. The brain tries to minimize prediction error between inputs and models about the causes of these inputs. Top-down connections of the brain convey predictions, whereas bottom-up connections transmit predictive errors. Expectations, from the perspective of Predictive Coding, refer to representations of causes encoded at each level. The same principles apply to interoception. The architecture of the insula allows for successively higher re-representations of interoceptive information, homeostatic significance, emotional salience and hedonic potential. The anterior insula integrates bottom-up interoceptive predictive error with high-level cortical predictions and sends interoceptive predictions to visceral systems. Generative models depend on actions. They actively map from hidden causes to sensory consequences. Perception corresponds to the inverse mapping, from sensations to hidden causes at higher levels of relational abstraction. The experience of body ownership fits a model of conscious selfhood based on interoceptive Predictive Coding underlying multisensorial integration. Predictive Coding supports a subject-centric view whereby actions not only lie at the basis of consciousness, but also provide the building elements of the subject itself. Neural correlates of consciousness represent what the subject is doing. The emergence of abstract phenomenal objects as guides to global adaptive behaviors favors the transition to phenomenological heuristics. Natural environments are characterized by low predictability, small sample sizes relative to the number of available cues, and interdependency among cues. Models based on heuristics are adaptive because predictive-error is difficult to calculate for global actions in an uncertain world.

Keywords: Active Inference, Bottom-Up, Brain Operations, Computational Hierarchy, Cortical Development, Embodied Abstractions, EPIC, Expectation, Feedback, Feedforward, Functional Connectivity, Generative Model, Heuristics, Interoception, Insular Cortex, Perception, Phenomenal Consciousness, Predictive Coding, Predictive Error, Predictions, Systematic Cortical Variation, Systems 1 and 2, Top-Down, Unified Stream of Consciousness, Winner-Take-All.

Paulo J. Negro

PREDICTIVE CODING SUPPORTS A SUBJECT-CENTRIC CONSCIOUSNESS

The brain hallucinates reality [109]. Perception is not a direct image of sensorial input. It is a hypothesis that explains sensorial information [110]. The ability to infer the cause of visual sensations as a face, for example, requires an internal generative model that integrates sensory evidence over time, *"more like a narrative or scenario, predicting sequences of events."* This generative model is known as "Predictive Coding".

The brain continuously attempts to minimize the prediction error between inputs and emerging models about the causes of these inputs[178] [108]. A generative model for perception necessarily includes multiple levels of description [110]. Expectations, from the perspective of Predictive Coding, refer to representations of causes encoded at each level.

Predictive Coding is consistent with Bayesian inference of hidden causes in the environment. Predictions originate from expectations at a level above. They are conveyed top-down[179]. Predictive errors are transmitted bottom-up by forward connections. They modify expectations in an iterative way[180].

Adams and colleagues [110] described how expectations encoded in the motor cortex can predict bodily sensory states and expected movement trajectories. They favor the term "intentions" instead of motor "commands". Intention is a more descriptive term because the dynamics of motor behavior are dictated not by precise prescription of muscular forces, but by the control of spinal reflexes responding to predictive signals.

The same principles have been applied to interoception. Feldman Barrett and Simmons introduced the term Embodied Predictive Interoceptive Coding (EPIC) to depict a model of active inference in interoception[181], in accordance to the expected flow of information across the laminar architecture of cortical connections [111].

Interoceptive-selective regions located in the dorsal-mid insula are characterized by granular cortex [103]. Their activity correlates with activity in the sensorimotor cortex proper (posterior insula and post-central cortex). The midventral insula also becomes active in interoception. However, resting state activation of the midventral insula correlates with the activation of the anterior insula, the subgenual anterior cingulate cortex and the temporoparietal cortex, indicating a mediating role in interoception and emotional regulation. The anterior and ventral insula is characterized by dysgranular cortex.

Prediction of sensory states by testable models allows the brain to efficiently maintain homeostasis[182]. Altogether, these observations indicate [102] that the anterior insula integrates bottom-up interoception prediction error with top-down high-level cortical predictions and sends descending interoceptive predictions to visceral systems.

Fig. (**5**) provides a brief display of cortical connection patterns [112 - 114] of feedforward and feedback pathways.

PREDICTIVE CODING IS CONSISTENT WITH BAYESIAN INFERENCE OF HIDDEN CAUSES IN THE ENVIRONMENT

Generative models depend on actions. They actively map from hidden causes to sensory consequences. Thus, perception corresponds to the inverse mapping from sensation to hidden causes [111] of higher levels of relational abstraction[183].

An elegant study of exteroceptive and interoceptive manipulation of the rubber hand illusion combined computer-generated augmented-reality with feedback of interoceptive cardiac information [116]. In this study, a "virtual rubber hand" in the subjects' visual field was modulated either in synchrony, or out of synchrony, with their actual heartbeat. The authors concluded that interoceptive signals directly affected the experience of virtual hand ownership *via* multisensory integration.

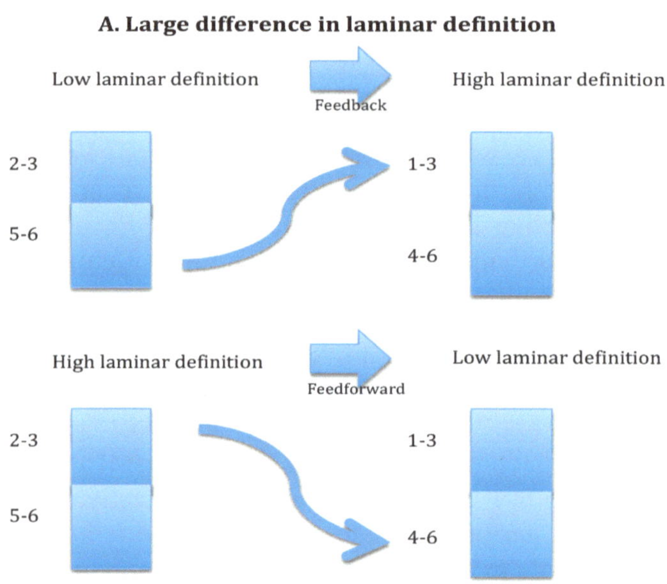

Fig. 5 contd.....

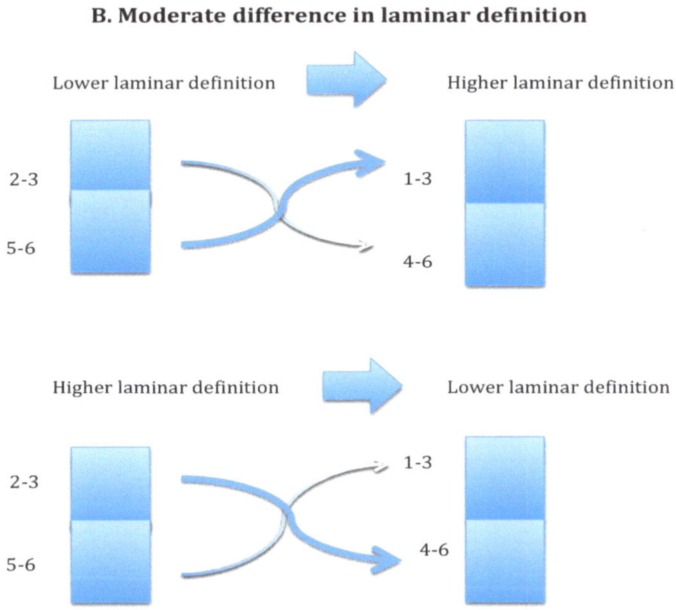

Fig. (5). Summary of predicted patterns of cortical connections [114]: 2-3 and 1-3 are superficial layers. 5-6 and 4-6 are deep layers. Reproduced with permission from Oxford University Press. A. Connections between cortices with large differences in laminar definition. B. Connections between cortices of moderate difference in laminar definition. Cortical limbic areas send feedback projections to lower level better-laminated areas. Feedback projections are seen as predictions driving cortical processing of information. These limbic areas are at a privileged position in the hierarchical cortical flow and may play a central role in providing a "workspace" for conscious experiences [115]. The paradigm *expectation.information* can be interpreted as *feedback.feedforward* or *prediction.predictive-error* and the realization of information characteristic of human consciousness as the strange loop:

'expectation.information'expectation.information

~

'prediction.predictive-error'prediction.predictive-error

The experience of body ownership constitutes a central aspect of selfhood. These findings fit a model of conscious selfhood based on interoceptive Predictive Coding[184]. Perceptual signals incorporated into multisensory predictive models update predictive models of body and self through the minimization of prediction errors[185]:

> *"We interpret our results in the light of emerging models of selfhood based on 'predictive processing' which focus existing multisensory integration accounts by positing that experiences of selfhood and body ownership are determined by the brain's estimation of the most likely interpretation of the ensemble of self-related signals, computed according to Bayesian principles"* (Fig. **6**).

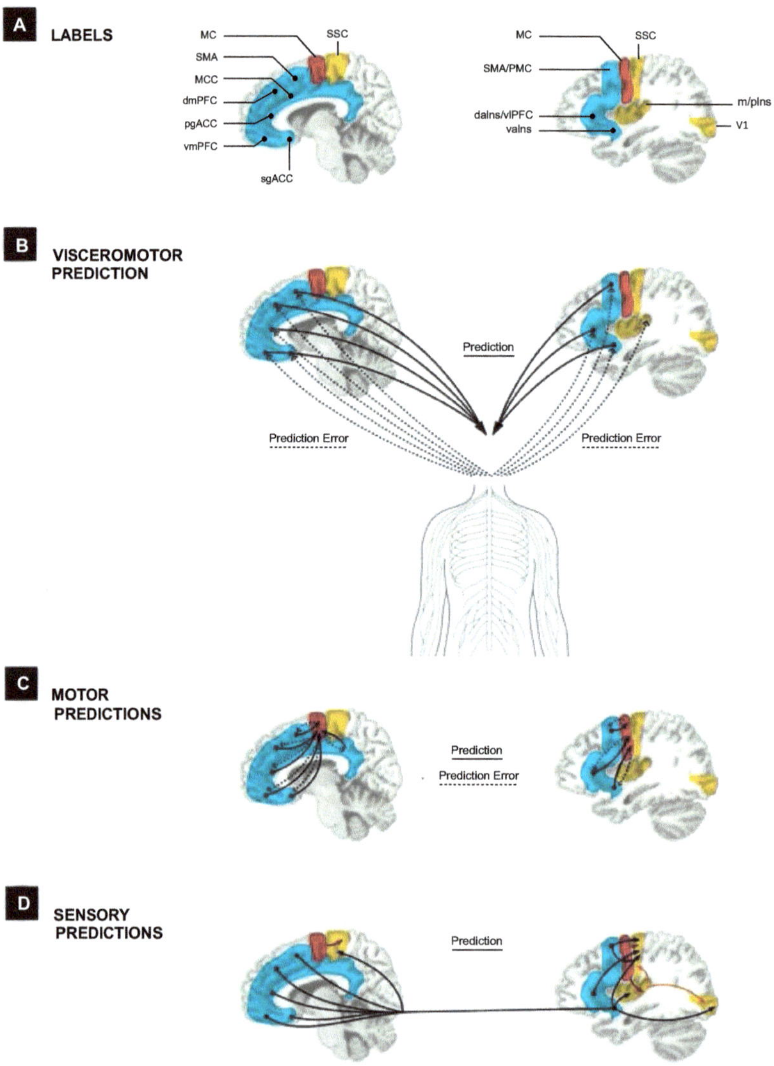

Fig. (6). Predictive coding in the human brain [117]. (A) Key limbic and paralimbic cortices (in blue) provide cortical control the body's internal milieu. Primary MC is depicted in red, and primary sensory regions are in yellow. (B) Limbic cortices initiate visceromotor predictions to the hypothalamus and brainstem nuclei to regulate the autonomic, neuroendocrine, and immune systems (solid lines). The incoming sensory inputs from the internal milieu of the body are carried along the vagus nerve and small diameter C fibers to limbic regions (dotted lines). (C) Efferent copies of visceromotor predictions are sent to MC as motor predictions (solid lines) and prediction errors are sent from MC to limbic cortices (dotted lines). (D) Sensory cortices receive sensory predictions from several sources. They receive efferent copies of visceromotor predictions and efferent copies of motor predictions. Sensory cortices with less well-developed lamination (*e.g.* primary interoceptive cortex) also send sensory predictions to cortices with better-developed granular architecture (*e.g.* in this figure, somatosensory and primary visual cortices)[186]. Reproduced with permission from Oxford University Press.

HOMO HEURISTICUS

Is it true that Bayesian reasoning underlies human behavior? [187] According to the work of Nobel Prize laureate Daniel Kahneman[188] [118], certainly not. Humans are poor Bayesian calculators. Their decisions rely less on statistical facts and more on heuristics. Heuristic methods involve simplifying rules that find a good enough solution for a given problem. In fact, Kahneman's research indicates how the reliance on heuristics is the source of bias in human beliefs and expectations. Biases or cognitive illusions are the rule rather than the exception. How can this be consistent with a predictive coding model hinging on Bayesian foundations?

Predictive Coding works well for cybernetic systems. It fits brain computation models because it optimizes motor action. All brain activity must be understood in light of action [78]. Nonetheless, Bayesian calculations are best suited for situations in which all relevant alternatives, consequences, and probability distributions are well known [119]. This may be suitable for brain computations that control movements, process incoming perceptual stimulation and build up neuronal assemblies representing abstract objects, but not necessarily for enacting adaptive global behaviors.

Bayesian computation models demand vast data space models to represent all possible probabilities and make exact calculations [120]. In contrast, sampling does not require knowledge of whole distributions and may thus provide an alternative and a good enough approximation to Bayesian inference. Active sampling is consistent with awareness as action. Bayesian sampling supports Friston & Stephan's [72] optimization of perception through selection of sensory samples. It may also explain the emergence of heuristics within the human mind.

Adaptive global behaviors, including the adaptive selection of phenomenal pre-representations[189], can be better achieved by heuristics than by complicated computational calculations[190] [121]. For example, a baseball player (or a dog) does not need complicated calculations to catch a ball. They can fix the gaze on the ball, start running and adjust running speed to keep the gaze angle or the image of the ball raising the same. We can see here how the emergence of phenomenal consciousness provides useful analogic solutions to difficult computational problems. Both for the baseball player and the dog the phenomenal constancy of a ball across various light situations, distance and angle supports adaptive action.

It is worthwhile to remind the reader of Stephen J Gould's [122] classic discussion about the origin of complex anatomical and functional structures such as the wings of insects. Proto-wings could have provided thermoregulation advantage to these animals. However, if these structures also provided aerodynamic advantages, such as gliding ability, there could have been a functional shift favoring the selection of wings with better aerodynamic characteristics that eventually allowed insects to fly. Notwithstanding current evolutionary theory discussions on the origin of insect wings, a functional transition from Bayesian calculation to a heuristics ecological reasoning can be expected.

Adaptive global behaviors are based on inferences stemming from sparse, not exhaustive observations. Despite Kahneman's focus on the shortcomings of heuristics and their role in cognitive illusions, heuristics can be phenomenally adaptive. In an unpredictable and poorly observable world, simple heuristics can very well decrease free-energy more efficiently than comprehensive Bayesian computations[191].

Phenomenal objects are generative synthetic constructs. Their adaptive value is not calculated inwardly, but by their performance as part of global actions. The performance of ecologically rational heuristics is best reflected

> *"not as a single number such as predictive accuracy, but as a learning curve revealing how bias and variance change as more observations become available" [121].*

Phenomenal objects serve to decrease perceptual variance and have a role in a form of ecological rationality. Predictive-error must be derived from the fitness value of the expectations that generate phenomenal objects. However, this is only possible through the interaction of the animal with its environment, which favors winner-take-all coalitions for phenomenal objects that match the environment. The GNW theory reports winner-take-all dynamics in the global integration of information. Winner-take-all global dynamics were not "designed" for consciousness but evolved from the dynamics of phenomenal objects.

CONSCIOUSNESS, FAST AND LOW

Kahneman's work was not focused on the origins of phenomenal experiences, but on how the human mind "thinks" about choices and calculations. According to him, the human mind operates two functional systems:

> *"System 1 operates automatically and quickly, with little effort and no sense of voluntary control. System 2 allocates attention to the effortful mental statistics that demand it, including complex computations. The operations of System 2 are often associated with the subjective experience of agency, choice, and concentration" [118].*

System 2 can be seen as the expression of explicit relational processes working primarily through conscious contents, such as choices, evaluation, comparison, grouping, categorization, serial ordering [118]. These events are made available simultaneously in the conscious field for the structuring and restructuring of novel relationships and troubleshooting function.

Both systems are associated with consciousness[192]. In the baseball player example, phenomenal awareness of the baseball demands very little effort and is experienced as automatic. The same may not hold true for the overall behavior necessary for catching the ball. An experienced baseball player may catch a falling ball with little perceived effort while a novice will require the conscious allocation of more cognitive effort to do the same.

Consciousness belongs to a hybrid system in which serial steps of conscious decision-making organize massive parallel non-conscious computation [46, 47]. As we will see later, the GNW theory is consistent with a model for consciousness characterized by winner-take-all global neuronal assemblies. The experience of a stream of consciousness is characterized by a winner-take-all dynamics within the serial steps that give rise to human consciousness. Such a process did not emerge by design. It evolved in the context of a conservative natural history. The same can be said about the origin of more abstract human cognitive heuristics and biases.

A parsimonious explanation would have all of the above share the same evolutionary origin in the winner-take-all heuristics of phenomenal objects. Binocular rivalry[193] serves as an example of winner-take-all mechanics. It represents an example of the mosaic of processes that characterize human consciousness. However, more importantly, it provides a clue to how awareness evolved from phenomenal objects into a unified single stream of consciousness by extension of a winner-take-all mechanics.

The emergence of abstract phenomenal objects might thus have favored the selection of robust informational systems. Such systems are less oriented towards modeling the past, and more towards performing well in a future that holds surprises, which decreases free-energy. Phenomenal objects are adaptive solutions supporting the origin of a mind that must make inferences from few observations and exploit adaptive biases.

Takashi Maeno [123] argued that phenomenal consciousness could not be separated from the evolutionary processes that enable episodic memory. He emphasized episodic memory as a model that

> *"memorizes cause-and-effect relationships that develop over time, [and]*

> *can be generated relatively easily by slightly modifying forward models that memorize simple cause-and-effect relationships."*

Maeno saw no adaptive advantage for phenomenal consciousness in organisms that have no episodic memory. The existence of conscious human beings with profound impairments of episodic memory due to bilateral hippocampal damage would seem to counter this idea [124]. However, the clinical finding does not truly refute Maeno's view. Episodic memory impairment is a *post hoc* clinical event in the context of a human evolutionary history already marked by the matching of episodic memory and phenomenal consciousness.

Nonetheless, his Passive Consciousness Model (PCM) can be considered consistent with some of the central ideas in this book. Maeno proposes that information constituting the mind is processed nonconsciously rather than by conscious systems, which are more passive recipients of such information processing. The PCM algorithm supports an internal model of the body and the external world, the matching between sensory input and expected consequences of behaviors, and the inverse modeling of predicted behaviors such as presented here.

Phenomenal objects can by themselves be adaptive as they support global adaptive actions. Late evolution of episodic memory might have allowed for the expansion of phenomenal objects as general cognitive strategies and, eventually, as the narratives that characterize human consciousness. According to Kahneman, nonconscious narratives underlie System 1 cognitive illusions. In contrast, a sense of voluntary control, attention, choice and concentration characterizes System 2 higher-level conscious experiences. System 2 highlights the link between action and agency that supports one of the fundamental ideas of this blueprint: action gives origin to the subject.

CONCLUSIONS

Generative models do not have an intelligent design bias. The subject has a central role in the integration of information because the subject gives origin to the action. Inverting the argument also yields a valid statement: action has a central role in the integration of information because it gives origin to the subject.

Predictive Coding is a top-down model based on actions or intentions. It supports a subject-centric view in which actions are not only the source of consciousness, but also the building elements of the subject itself.

Predictive coding lends support to a theory of consciousness based on the active realization of information: active expectation. From this perspective, qualia become "beliefs"[194] about causes and states in the world. The brain constantly updates these beliefs by evaluating expectations and predictive-errors[195].

The insular cortex activation[196] during interoception is consistent with the participation of the circuitry in computations under the formal constraints implied by Predictive Coding. Thus, Predictive Coding has a connection with mechanisms of awareness.

However, generation of pre-representational and representational phenomenal objects may characterize a transition from predictive coding computations into the system dynamics that underlie simple heuristics. Natural environments are characterized by low predictability, small sample sizes in comparison to the number of available cues, and interdependency among cues. Predictive coding computations fail when predictive-error is difficult to compute. Such difficulties arise in an uncertain world where probability distributions are not known. Phenomenal objects decrease perceptual variance and provide a form of ecological rationality with a faster learning curve. Predictive-error must be derived from the fitness value of the expectations that generate phenomenal objects, and this only occurs through the interactions of an organism with the environment.

NOTES

[178] Gray [309] suggested that the contents of consciousness were the output of a comparator matching the current state of the organism's perceptions with predicted states.
[179] Top-down: feedback or backward connection.
[180] Expected precision refers to the relative confidence in descending predictions and ascending prediction errors [110]: a "prediction" is considered to be generated from an "expectation" by nonlinear transformation. A single expectation can generate multiple predictions in different contexts.
[181] As reviewed, the posterior and middle parts of the insula have substantial reciprocal connections with the somatosensory cortex, and receive nociceptive, thermal and afferent projections [103]. The anterior portions of the insula are strongly connected to prefrontal and limbic structures, such as the anterior cingulate and orbitofrontal cortices, the amygdala and the ventral striatum. The anterior insula participates in interoceptive awareness and processing emotional stimuli, particularly negative valenced stimuli. Negative valenced: disgust, guilt, perception of fear-related pictures, retrieval of sad memories, as reviewed by Simmons and colleagues [103]. The left anterior insular cortex (AIC) is activated in all valence emotions [102]. Although the AIC is commonly coactivated with the anterior cingulate cortex (ACC), the AIC can dissociates from the ACC in emotional awareness. According to Gu and colleagues [102], the few studied patients with AIC damage showed that the AIC is necessary for explicit empathetic pain process. Both stimulus-driven and top-down information were impaired in these patients. The same was not observed in patients with lesions of the anterior cingulate cortex (ACC). Insular lesions do not necessarily abolish interoceptive awareness and emotional

awareness. This is interpreted as interoceptive and emotional awareness not being processed by a single neural substrate, but by a distributed processing by the brainstem and thalamic nuclei in emotional awareness and frontotemporal processing in self-recognition tasks. Furthermore, bilateral insular damage can be capricious and impair, but not abolish, interoception. This architecture is consistent with successively higher re-representations of interoceptive information, homeostatic significance, emotional salience and hedonic potential. It positions the insula as a hub of multiple systems underlying sensation, emotion, and cognition.

[182] Generative models efficiently maintain homeostasis because they minimize the (information dispersion) entropy of their interoceptive and exteroceptive states [112].

[183] Conditional expectations [111] are top-down predictions. Predictive errors are sent bottom-up to update these conditional expectations. There is an implied functional asymmetry that allocates nonlinear functions to top-down and lateral predictions, and linear functions to the feedforward signals of prediction errors.

[184] *"The predictive coding views adapts self-perception to a "Bayesian brain" framework, which describes perception as the process of inference about the most likely causes of sensory input. On this view, self-representation consists in multi-level (hierarchical) generative (i.e.* predictive) models of the cause of those sensory signals which are deemed most likely to be me" [116].

[185] The minimization of prediction error is crucial to update posterior probabilities and induce changes of prior probabilities. In this virtual reality study, statistical correlations between highly weighted sensory signals (vision, touch) and interoceptive signals (a subtle change of the color of the virtual hand with heartbeats) overcame prediction errors in proprioception.

[186] Comparisons between prediction signals and ascending sensory input results in prediction error that is available to update the brain's internal model. In this way, prediction errors are learning signals and therefore adjust subsequent predictions. For simplicity, only primary visual, interoceptive and somatosensory cortices are shown; subcortical regions are not shown. Prediction errors are not depicted in panel D;sgACC, subgenual anterior cingulate cortex; vmPFC, ventromedial prefrontal cortex; pgACC, pregenual anterior cingulate cortex; dmPFC, dorsomedial prefrontal cortex; MCC, midcingulate cortex; valns, ventral anterior insula; dalns, dorsal anterior insula and includes ventrolateral prefrontal cortex; SMA, supplementary motor area; PMC, premotor cortex m/pIns, mid/posterior insula (primary interoceptive cortex); SSC, somatosensory cortex; V1, primary visual cortex; and MC, motor cortex.

[187]"Homo heuristicus": a quote from Gigerenzer and Brighton [121].

[188] Daniel Kahneman in collaboration with Amos Tversky developed the scientific field of judgment and decision-making. Their work culminated in the development of the Prospect Theory, which was profoundly influential in the field of behavioral economics. Readers can refer to their original publication "Prospect Theory: an analysis of decision under risk" [206]. As well as to his assessable popular science book "Thinking, fast and slow" [119]. Prospect Theory describes how choices are made between alternatives that involve risk. Value is assigned to gains and losses, rather than final assets. These choices are based on certain heuristics assessing the potential value of losses and gains. Cognitive illusions can arise from these human heuristics, such as base-rate neglect, conjunction

fallacy, confirmation bias, among others.

[189] The reader should refer to the work of Jean-Pierre Changeux [68]. He sees the onset of agreement between brain states and external reality to antecede the appearance of *Homo sapiens*. According to Chageux, knowledge acquisition results from the selection of pre-representation, defined as *"dynamic, spontaneous, and transient states of activity arising from te recombination of preexisting populations of neurons"*.

[190] The Adaptive Behavior and Cognition group at the Max Planck Institute for Human Development addressed heuristics from a different perspective [119]. Heuristics are processes that ignore information and enable fast decisions. This can be seen as a manner to save energy by trading accuracy with effort. However, Bayesian calculations work best at "small worlds" where all relevant alternatives and probabilistic distributions are well known. This is not the case in a "large world" where probabilities are either not known or knowable. In this case, matching of heuristics with the environmental structure is more relevant to outcomes: *"a heuristic is ecologically rational to the degree that it is adapted to the structure of an environment."* It may actually make predictions with less information, faster and more accurately.

[191] *"The ideal is to have an infinitely flexible system of abstract representations to ensure zero bias, so that whatever structure the world has, it can be reproduced perfectly. As the content of samples of observations in this world will vary, the induced representations are likely to be different, and this creates variance. Such a mind works best with large samples of observations and in a world that is relatively stable. Yet because this mind has no bias and must choose from an infinite space of representations, it is likely to require resource-intensive cognitive processing. This kind of design suggests general-purpose processing strategies such as exemplar models and neuronal networks as models of cognition. ...To achieve low prediction error on a broad class of problems, to accommodate both linear and nonlinear patterns not known in advance, and accidental noise patterns, a low bias model requires an increase in variance. It is prone to make inaccurate predictions. That's the bias-variance dilemma that shows the reason the mind can be "better off with an adaptive tool box of biased, specialized heuristics ...to control variance, one must abandon the ideal of general-purpose inductive inference and instead consider, to one degree or another, specialization"* [119].

[192]Crick and Koch already emphasized a distinction between fast and slow awareness in 1990 [306].

[193] Binocular rivalry: the alternating visual perception instead of superimposed images when two different images are presented to each eye.

[194] Sensory inputs constrain estimates of prior probabilities (from past experiences) and generate posterior probabilities that serve as predictions of presenting sensory inputs [106]. The brain forms neural representations based on previous experiences, anticipates sensory information, which implements as predictions propagating across cortical regions. Predictions (prior probabilities) serve as hypotheses about the world that can be tested against sensory information arriving at the brain.

[195] As per Adams and colleagues [110]: *"there is no explicit physiological categorization of the notional expectation, error and precision units, but we make the provisional assumption that all three are represented by pyramidal cells or by excitatory, spiny stellate cells in [cortical] layer 4"*.

[196] If it is assumed that agranular and dysgranular regions of the anterior and midventral insula are the source of top-down predictions and if it is confirmed that midposterior insula activation is associated with interoceptive experiences, information should be realized as interoception at the superficial layers of the midposterior primary insular cortex. The chapter addressing the role of the right fronto-orbital cortex (FOC) and the piriform cortex (PC) in conscious olfaction parallels this issue.

CHAPTER 14

Cortical Signals of Consciousness

Abstract: Late extracellular signals underscore some of the neural signatures of consciousness, which seem to depend on long-range feedback processing. The Visual Awareness Negativity and P300, which denote a variation in awareness, correlate with consciousness and originate from late signals. Mid- and long-distance information sharing provides long-range cortical communication, and also represents a neural correlate of consciousness in humans. In granular sensory cortices, superficial and deep pyramidal cells respectively process feedforward and feedback information, apparently as parallel streams. There is evidence identifying Layer-5 neurons as the main cortical column integrators of information. Layer-5 pyramidal neurons might function as coincidence detectors by signaling feedforward and feedback interactions with a high-frequency spike burst triggered by calcium influx. These spike bursts may give origin to the late extracellular signals observed at superficial cortical layers. The high-bursting model of coincidence detection fits the concept of realization of information: *'expectation.information'expectation.information* can be re-expressed as *'feedback.feedforward'feedback.feedforward.*

Keywords: Beta Oscillation, Calcium-Spike, Cerebral Cortex, Coincidence Detector, EEG Integration and Differentiation, Feedforward and Feedback Streams, *'feedback.feedforward'feedback.feedforward*, Gamma Oscillation, High-Bursting Model, Information Sharing, Local Field Potential, Long-Range Connections, Layer 1 Inhibition, Layer 5, Neural Correlates of Consciousness, Predictive-Error Suppression, Pyramidal Cells, P300, Superficial and Deep Cortical Layers, Signals of Consciousness, Visual Awareness Negativity.

A BRIEF REVIEW OF THE PHYSIOLOGY

In granular[197] sensory cortices, superficial and deep pyramidal cells respectively process feedforward and feedback. Although extrinsic cortico-cortical connections have an excitatory glutamatergic basis, their connection to superficial layer 1[198]provides an important route for selective feedback inhibition. Layer 1 effectively allows predictive signals from higher areas to deliver controlling information to lower cortical areas[199]. These connections suppress prediction error firing.

Paulo J. Negro

Although interconnected, superficial and deep layers seem to process parallel streams of information[200]. Maier and colleagues identified two distinct zones of local field potential (LFP) activity corresponding to superficial and deep layers in the primary visual cortex (V1) in macaque monkeys[201] [125]. A sharp transition at the bottom of layer 4 divides these functional domains. The signals were highly coherent within each laminar zone, but not between zones[202]. The functional dissociation between information traveling top-down and bottom-up was verified by Fontolan and colleagues [126], as discussed later.

SIGNALS OF CONSCIOUSNESS IN CORTICAL LAYER 5

Anastassiou and Shai pointed out measurable brain signals that correlated with consciousness [127]. Late extracellular signals underscored some of these neural signatures of consciousness, which seem to depend on long-range feedback processing[203] (Table **2**).

Table 2. Cortical brain signals correlated with consciousness [127].

LATE EXTRACELLULAR SIGNALS
The Visual Awareness Negativity[204] and the p300[205]: these signals denote a variation in awareness mediated by a late signal originating in the upper layers of the neocortex. Authors disagree as to whether these signals relate to conscious, pre-conscious or post-conscious events.
DISTRIBUTED INFORMATION SHARING
Sustained and widely distributed voltage deflections, increased of beta synchrony (12-20Hz) and gamma power (30-60Hz): these signals reveal a widely distributed cortical activity associated with conscious perception. (Activity restricted to primary sensory areas is not associated with conscious perception).
LONG-RANGE FEEDBACK CONNECTIONS
The amount of integration and differentiation distributed across the EEG quantify the “amount” of consciousness in patients. Mid- and long-distance distributed information sharing indicates long-range cortical communication. It is considered a neural correlate of consciousness in humans.

Shai and colleagues reported a single-cell mechanism for spike bursting based on dendritic calcium-spike of pyramidal neurons[206] [128]. Because calcium-spike may be the origin of late extracellular signals as a relatively slow and large-amplitude signal in the superficial layers, it may represent a neural correlate of consciousness. A calcium-spike can integrate with back-propagating action potentials to trigger a spike burst in the body (or soma) of the pyramidal cell.

According to Shai and coworkers, layer 5 neurons are the main integrators in the cortical columns projecting to other cortical and subcortical areas [128]. These cells have dendrites across all cortical layers, receiving both excitatory and local inhibitory stimulation[207].

They used a multi-compartmental model to study the capability of layer 5 pyramidal neurons to detect coincidences between their basal and apical tufts. They simulated the interaction of dendritic and somatic excitatory signals in a single pyramidal cell model. In the simulation, the coincident input of perisomatic and apical stimulation generated bursts of high frequency action potentials in the cellular body. However, when separated, the same perisomatic or apical stimulation did not yield the same effect[208].

L1 is a sparse layer, more than 90% composed of synapses from long-range input [128]. In Shai's model, feedforward connections innervate the basal dendrites of layer 5 with excitatory synapses, while feedback connections innervate pyramidal dendritic tufts at layer 1. The data suggests that L5 pyramidal neurons can function as a "coincidence detector", bursting with concurrent basal and apical dendrites stimulation, matching feedforward and feedback signaling.

According to these authors, feedforward and feedback interaction is signaled by a high-frequency spike burst of the pyramidal cells. This mechanism can underlie the association of information from multiple cortical areas necessary for specific cortical signals, differently from pure feedforward stimulation.

CONCLUSIONS

The high bursting model of pyramidal cells flags detection of coincidences from feedforward and feedback streams. It supports the concept of realization of information. However, Predictive Coding requires as revision of the *information.expectation* paradigm.

Original blueprint model:

'information.expectation'information.expectation

feedforward.feedback'feedforward.feedback

As Predictive Coding is based on actions or intentions, the strange loop should be updated as:

expectation.information'expectation.information

~

'feedback.feedforward'feedback.feedforward

The subject-centric action-based model of consciousness presented in this blueprint is thus noted as the strange loop:

'expect.info'expect.info

NOTES

[197] A granular cortex is organized into an external granular layer (II) and an internal granular layer (IV). A dysgranular cortex has fewer granule cells in a single layer and an agranular cortex has no granule cells.

[198] Layer 1 was historically called acellular due to paucity of neurons. It contains 0.5% of a cortical column cells. Its cells are almost all inhibitory, strongly connected with each other, and the source of strong monosynaptic inhibition to layers 2/3, which pyramidal cells apical dendrites extend to layer 1 [112, 113], with the bottom of layer 1 and top of layers 2/3 the inhibitory fulcrum of the column.

[199] Feedforward signals are theoretically and empirically excitatory. Activity in higher cortical regions depend on inputs from lower areas that establish their receptive field properties, as reviewed by Bastos and colleagues [112], through cortico-cortical and transthalamic pathways.

[200] Layers 2/3 express gamma oscillations when coexpressed with beta oscillations in layer 5 [113,310].

[201] They used a linear microelectrode array to measure LFP signals at multiple evenly spaced positions throughout the cortex.

[202] They found increased LFP gamma power in superficial layers in comparison to deep layers, and the reverse pattern in the alpha and beta ranges. Building on this previous work, Bastos and colleagues hypothesized that different high and low frequencies mediate feedforward and feedback signals. In this model, afferent information enters layer 4 and is conveyed to layers 2/3, which project to the next higher cortical region, after interacting with layer 1. This seems the mechanism of moving forward gamma oscillations through cortical regions [113]. In their canonical predictive coding microcircuit, expectations about causes and states are assigned to interneurons in the supragranulary layers, which are passed to infragranulary layers to encode local level prediction errors. Superficial pyramidal cells encode prediction errors sent forward to the next hierarchical level. Infragranulary excitatory interneurons generate feedback to superficial local layers. Infragranulary pyramidal cells send feedback predictions to the lower hierarchical cortex. According to the Fourier transformation of the mathematical equations of Bastos and colleagues [112], high frequencies should be attenuated as passing from superficial to deep cells. *"There is nothing mysterious about this attenuation – it is a simple consequence of the fact that conditional expectations accumulate prediction errors, thereby suppressing high-frequency fluctuation to produce smooth estimates of hidden causes. This smoothing – inherent to Bayesian filtering – leads to an asymmetry in frequency content of superficial and deep cells: for example, superficial cells should express more gamma relative to beta, and deep cells should express more beta relative to gamma."*

[203] According to Anastassiou & Shai [127], multiple studies showed interruption of perception by TMS over V1 (Visual area 1) both in the feedforward and feedback time frames. Single cell studies show early firing of cells in a general category (*e.g.* face) and later firing in a specific identity. Backwards masking studies (a target stimulus is masked is followed by a masking stimulus ~50ms later) showing no awareness of the target stimulus with masking indicate that disruption of feedback, not feedforward processing is associated with loss of awareness.

[204] As reviewed by Anastassiou & Shai [127]: in the 1960's EEG human studies comparing the averaged extracellular signature in simple detection tasks (visual and auditory) showed a significant difference in the amplitude of the negative wave detected 160ms after the signal onset: the amplitude of the wave correlated positively with perception.
[205] P300: also known as p3b or late potential.
[206] From Shai and colleagues [128]: the apical dendrites of pyramidal cells at L2/3 or L1 contains a high density of voltage-dependent calcium-channels which, in combination with other nonlinear channels, allow a relative slow all-or-nothing depolarization called "calcium-spike". Calcium-spikes can depolarize the pyramidal cell for a long period of time. These signals can be controlled by superficial layers inputs, including long-range feedback connections.
[207] Their work established active properties of L5 pyramidal dendrites (L5 pyramidal dendrites properties of mouse V1 area).
[208] More of this will be discussed in the section addressing general anesthetics and consciousness.

CHAPTER 15

Olfactory Awareness

Abstract: The olfactory system provides an especially propitious model for the study of neuronal correlates of consciousness. Olfactory signals mostly bypass thalamic first-order relay neurons and directly affect the olfactory cortex. In the olfactory cortex, the sparse distribution of odor-induced activity and the selective odor tuning of individual pyramidal cells resembles the hippocampus place field and place cells. In mammals, the olfactory cortex gave origin to the isocortex. The realization of information in this region may play the role of a heuristic process for the realization of information in other cortices. Binding of information in the olfactory cortex results from specific focal signaling in the context of widespread activation. Odorants bind to odorant receptors (OR) on the surface of olfactory sensory cells and activate neural maps in the olfactory bulb (OB). This process happens through the synchronization of localized tufts cells and the wide activation of mitral cells, and it fits with the *information.expectation* model of realization of information. The OB depicts odorant characteristics in spatiotemporal patterns. The piriform cortex (PC), considered the primary olfactory cortex, generates pattern recognition beyond the OB maps. Much like the hippocampus, PC oscillatory activity establishes cellular synchronicity thorough the ensemble and across the oscillatory phase. The posterior PC integrates structural codes into more highly processed representations of odor qualities, considered odor-object categories. The existence of odor-objects is consistent with the manipulation of information as a focal aspect of subjective experiences. Active mapmaking realizes information that transcends into qualia. The process of mapmaking represents a synthetic *de novo* phenomenon with a relational origin. Olfactory experiences are abstracted from the binding of specific molecules to olfactory receptors into structure-based categories, and further on into odor-object categories. Transitions happen under feedback input, in agreement with Predictive Coding. The overall process parallels the hierarchy of abstractions instantiated in the insula. The transition of external information into organism-centric encoding supports the idea of qualia as conveyors of adaptive information. The synthesis of low dimensional percepts supports the transition from the Bayesian calculations of Predictive Coding to phenomenal heuristics.

Keywords: Dimensional Reduction in Phenomenal Objects, Gamma Oscillations Realization of Information, Mitral Cells, Odorant Receptor, Odor-Object, Olfactory-Hippocampal Axis, Olfactory System, Olfactory Bulb, Orbitofrontal Cortex, Organism-Centric Encoding, Oscillatory Coupling, Phenomenal Heuristics, Piriform Cortex, Predictive Coding Transition, Principal Components Analysis of Perceived Odors, Realization of Information Across Abstraction Levels, Synthesis of Qualia, Tuft Cells.

Paulo J. Negro

Phylogenetically, olfaction is possibly the oldest exteroceptive[209] sense. The olfactory system presents several advantages over other sensory modalities in terms of studying the neural correlates of consciousness [129]. Unlike other afferent sensory signals, olfactory signals mostly bypass thalamic first-order relay neurons and directly affect the ipsilateral olfactory (piriform) cortex. The olfactory system provides a good model for the study of neural correlates of consciousness because of its neuroanatomy, phenomenology, cognitive properties, and oscillatory properties.

A BRIEF REVIEW OF THE PHYSIOLOGY

Conscious olfactory perception occurs during discrete respirations (sniffs). Olfactory sensory neurons (OSN) begin activation at the rising phase of inhalation, continue through the rest of inhalation, and decline during the exhalation phase. The inhalation phase is the active phase, during which the central olfactory system processes external odor information in real time. Exhalation isolates the central olfactory system from external stimulation, particularly in the latter part of a long exhalation[210] [130].

During the slow respiration of sleep, the olfactory cortex and higher association areas are largely isolated from external odors due to a behavioral state-dependent sensory gating[211]. Mori and Manabe [130] used the term *"translation-movement theta sniffs"* (t-theta sniffs) in contrast to *"translation-movement theta oscillation"* to address the olfactory oscillations during movement [130]. The olfactory cortex and high olfactory centers also see oscillatory activity during the exhalation phase up to the beginning of the following inhalation. During this period, the central olfactory system remains off-line and disconnected from external odors.

The olfactory bulb (OB) is the first relay of odor information in the central nervous system. It receives sensory input from OSN at the olfactory epithelium in the nose[212]. Anatomically, the OB is constituted of multiple glomeruli.

Each OSN expresses only one odorant receptor and projects its axon to a single glomerulus at the OB. Axons of olfactory neurons expressing the same receptor converge into the same glomerulus. Therefore, both OSNs and glomeruli follow the same rule: *"one cell-one receptor"* and *"one-glomerulus-one receptor"*.

FROM ODOR RECEPTORS TO OSCILLATIONS IN THE OLFACTORY BULB

According to current receptor theory, binding of the odorant to the odor receptor triggers odor perception[213]. Each OR receptor can recognize structurally similar odorants, and subgroups can recognize specific odorants. The combinatory activation of OR receptors allows for the discrimination of different odorants [131, 132].

The physiologic matrix of odorant-odorant receptor pairs connecting the animal with its environment differs from the olfactory map created in the olfactory bulb[214]. Afferent axons from the olfactory epithelium connect to the dendrites of the principal cells in the glomeruli, which are divided in mitral and tufted[215] cells.

Tufted cells are activated early in the sniff cycle[216] by low odor concentrations. They fire high-frequency burst discharges. Mitral cells are activated by higher odor concentrations. They fire lower-frequency burst discharges beginning in the transition from inhalation to exhalation and continuing through part of the period when the olfactory bulb and cortices are off-line [132].

Tuft cell circuits generate sniff rhythm-paced early-onset fast gamma oscillations. Each pair[217] of glomeruli responds to a range of odorants sharing combinations of molecular characteristics. The activation of glomeruli combinations yields a basic neural map representing the molecular features of odorants [131 - 133].

EARLY AND LATE OLFACTORY GAMMA OSCILLATIONS ARE CONSISTENT WITH *EXPECTATION.INFORMATION*

Tufted cell circuits largely generate the early onset fast gamma oscillations[218] of the inhalation phase. The late onset slower gamma oscillations of the inhalation to exhalation transition are generated by mitral cell circuits [132].

Fast gamma synchronization of tufted cell circuits among relatively short regions is followed by slower gamma synchronization of mitral cell circuits among larger areas of the olfactory bulb. Furthermore, tufted cells send odor information to more restricted areas of the olfactory cortex. Mitral cell signals are more widely distributed.

Circuits often project information in specific fashion, much as observed with the early onset, rapidly synchronized signal of a predator odor. The difficulty lies in attempting to understand the role of widely dispersed mitral cell axons, projected to almost all areas of the olfactory cortex including sister mitral cells that send odorant information to nearly all pyramidal cells in the piriform cortex (PC)[219] [132]:

> *"...although individual pyramidal cells receive only weak inputs from individual mitral cells, nearly simultaneous arrival of the synchronized inputs from many mitral cells may effectively summate their excitatory postsynaptic potentials and thereby strongly modulate pyramidal cells*

activity in synchrony with the slow gamma oscillation inputs. We hypothesize that the gamma-synchronized coincident inputs from many mitral cells coordinate the response timing of pyramidal cells that are spatially distributed across whole parts of the piriform cortex over a sustained time window. This mitral cell-induced synchronized activity of pyramidal cells across whole parts of the piriform cortex may have a key role in 'binding' together the spike activities of numerous coactivated pyramidal cells with different odor tuning specificity..."

The binding of information in the olfactory cortex therefore results from specific focal signaling in the context of widespread activation[220]. This process fits with the model *information.expectation* or, in other words, with the "reading" of olfactory bulb maps as realization of olfactory information[221]. It resembles the realization of information in hippocampal place cells[222].

Binding depends on synchronized activity[223] [134 - 137], but synchronized activity carries the risk of false bindings when objects share the same features [136, 137]. The realization of information through *information.expectation* loops points towards a solution to this problem. Such solution rests on the enhancement of synchronization leading to selection of neuronal assemblies based on the combination of bottom-up and top-down effects [138].

INFORMATION FLOW ACROSS THE PIRIFORM CORTEX AND THE EMERGENCE OF "ODOR-OBJECTS"

The piriform cortex (PC), divided into anterior (APC) and posterior (PPC) piriform cortices, is the largest region of the olfactory cortex[224]. A detailed analysis of PC anatomy would far exceed the scope of this book. It suffices to point out that such anatomy fits nicely with the predictive coding model.

Olfactory sensory neurons (OSN) converge onto the glomeruli in the olfactory bulb, generating a spatial pattern of activity in second-order neurons. These odorant-specific olfactory bulb patterns are projected to the olfactory cortical areas[225]. OSN projections are always ipsilateral, *i.e.* within the same side of the brain. Cortical neurons project and access bilateral information. The PC, a paleocortex[226] composed of three layers[227] [139], forms the largest region in the olfactory cortex.

Feedforward projections from tufted and mitral cells connect at layer 1a. On the other hand, deep recurrent association fibers initiate within each area of the PC and extend back to their own PC area and other areas, preferentially to layer III of the whole PC[228]. Feedforward and feedback connections become active during different periods of the sniff cycle[229]. In addition, pyramidal cells within the PC have feedforward extensions into high association areas, including the orbitofrontal cortex (OFC), thalamus, hypothalamus and ventral striatum. The OFC receives multimodal, sensory information including gustatory and olfactory signals[230].

Individual PC neurons respond to specific combinations of odorants:

> *"...the sparse distribution of the odor-induced activity and the selective odor tuning of individual pyramidal cells resemble those of the hippocampus[231] ('sparse distribution of CA1 pyramidal cells (place cells) that fire at a particular place and individual pyramidal cells have the property of well-tuned place cell')" [133].*

The PC resembles the hippocampus. It is a combinatorial, auto-associative cortex that generates content-addressable memory [139]. The olfactory-hippocampal axis forms a key network for spatial navigation and orientation in mammals. Most importantly, in mammals it gave origin to the isocortex [140]. The realization of information in these regions may thus play the role of a heuristic process for the realization of information in other cortices [232].

In humans, the olfactory epithelium (OE)[233] acquires odorant information and transmits it through the cribriform plate to the OB in the brain. The OB projects feedforward information to the olfactory cortices. The human PC is considered the primary olfactory cortex [234], and the human OFC the main secondary olfactory cortex [141, 142].

The first PET [235] study on olfaction revealed that a simple odor presentation causes bilateral activation of the PCs and activation of the right OFC [143]. The authors identified the PC as the primary olfactory cortex, and the right OFC as the secondary olfactory cortex, involved in higher-order odor processing. Kjelvik and colleagues found that odor identification in humans activates the entorhinal cortex, the right hippocampus, and secondary olfactory structures, including the OFC[236] [144].

In humans, an increase in fMRI bold[237] signal in the entorhinal cortex and the hippocampus coincided with correct olfactory identification. Identified and unidentified odors increased bold signal in the right OFC and the PC, but identified odors caused a greater increase than unidentified odors[238]. However, not all authors agree with a role for the PC as the actual primary olfactory cortex[239] [145 - 147]. At any rate, there is reason to see the right OFC[240] as participating in the realization of information subsequently realized in the integrated global workspace of consciousness [241].

In summary, the OB depicts odorant characteristics in spatiotemporal patterns based on glomerular activation. The PC transitions these spatiotemporal into distributed hippocampal-like ensembles of olfactory cortical pyramidal cells. The olfactory cortex generates a pattern recognition network beyond the structural odorant characteristics out of OB information. This pattern recognition is influenced by the past history of activity in distributed cortical ensembles[242].

OB afferents connect preferentially with the APC, while PPC afferents predominantly originate from the associational fibers of other regions. The APC receives structure-based information arising from the OB[243] and likely *"provides the biological substrate (encodes) for odorant structure codes"* [148]. The PPC integrates structural codes into more highly processed representations of odor qualities. Consistently with the predictive coding paradigm, odor quality is created under the effect of feedback afferents from other association regions to the PPC[244] (Table **3**).

Table 3. From gamma oscillation to functional states [149].

GAMMA OSCILLATIONS Early-onset fast and late-onset slow APC gamma oscillations phase-couple with OB fast and slow gamma oscillations. Slow gamma oscillations at the OFC at times phase couple with the APC and the OB. Phase coupling of APC and OFC gamma oscillations can occur both in the inhalation and exhalation, when the olfactory system is off-line. This indicates both bottom-up and top-down connectivity.
ODOR PERCEPTION To perceive an odor, the brain needs to know the specific combination of gamma oscillators co-activated by the odor inhalation. Perception is dependent on a sustained background of oscillations for the binding of distributed neuronal activities.
FUNCTIONAL STATES Large-scale gamma oscillation couplings across thalamocortical and corticothalamic regions generate functional states that characterize conscious processing. In comparison: thalamocortical projection neurons are classified in core and matrix neurons. Core neurons project topographically to the middle neocortical layers in an area-specific manner. Matrix neurons project diffusely to the superficial layers of the neocortex. The functional roles of tufted and mitral cells resemble the general pattern of core and matrix neurons.

The PC promotes the transition from odorant structure representations into *"odor-object categories, akin to visual object forms in human central temporal cortex"*. Spatially distributed ensemble activity in the human PPC is shown to coincide with perceptual ratings of odor quality, but these fMRI patterns are not found in the APC, amygdala or OFC[245] [150].

Thus, the PPC, and to some degree the APC, organize information in odor-object categories that are consistent with the manipulation of information as focal aspects of subjective experiences. Afferents from other regions, such as the OFC and the hippocampus establish higher-level categories that support semantic associations, sensory context, and perceptual learning[246] [148] .

THE MATCHING OF OLFACTORY AND INTEROCEPTIVE HIERARCHY OF ABSTRACTIONS

Olfactory receptors bind to odorants that convey specific molecular patterns. These molecular feature maps do not exist in the environment, they are abstracted from the binding of specific molecules to the ORs[247][130 - 133]. As seen above, structure-based categories [151] (APC) are further abstracted in odor-object categories (PPC) under increased feedback input from higher olfactory processing [131 - 152] [248] including sensorial context, episodic memory, and semantic associations[249]. The feedback input can be conceivably established under Predictive Coding[250] (Fig. **7**).

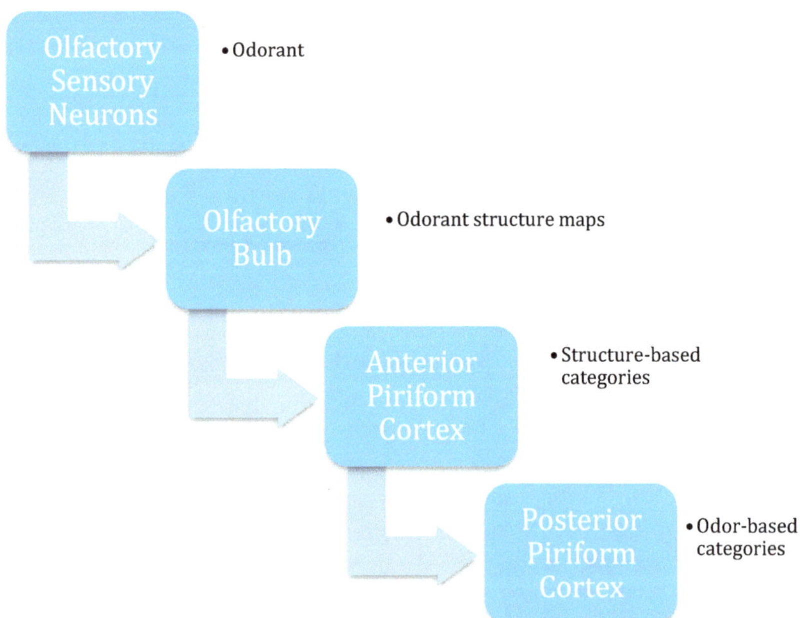

Fig. (7). Hierarchical abstraction of information from molecular patterns to structure-based abstractions to odor-object abstractions. This pattern resembles successively higher re-representations of interoceptive information, homeostatic significance, and salience in the insula.

Information that eventually transcends into qualia (*i.e.*, focal aspects of awareness) is realized from active mapmaking, instead of corresponding to specific odorant receptors binding to specific environmental odorants[251]. Olfactory qualia generated by higher centers do not necessarily correspond to anything physical in the environment. These are *de novo* phenomena with a relational origin [71] based on the organism history, as indicated by the heavy feedback fibers from higher associative stations. This "best guess" adaptive action about environmental information is conveyed as "qualia"[252].

THE SYNTHESIS OF SMELL IS CONSISTENT WITH QUALIA AS INFORMATION

Weiss and co-workers reanalyzed the dataset "Atlas of Odor Character Profiles" collected in the 1980's, containing numerical scores of 150 experts rating 138 monomolecular odorants along 146 verbal descriptors [141]. They hypothesized that the functional dimensionality of the atlas should be lower than the 146 dimensions generated by the verbal descriptors. In fact, a principal components analysis of the dataset demonstrated that a small number of primary components[253] (PrCs) explained most of the variance (Fig. **8**).

The first PrC (PrC1) explained about 30% of the variance. The *"low-dimensional olfactory percept [that] dominated one perceptual axis"* was odor pleasantness. Similar conclusions emerged from other experiments in rodents, and even from the identification of odorant valence as a coding axis represented at the output level of the antennal lobe of the fly brain [141,153-156].

Weiss and colleagues analyzed the functional dimensionality of ~1,600 chemical descriptors for ~1,500 odorant molecules [141]. They found that a small number of PrCs predicted most of the observed variance. Odorant structure (molecular size and compression) explained approximately 30% of the variance. The authors named this PrC "compactness". Compactness of odorant structure correlated with pleasantness of odor perception.

The same authors also studied the functional dimensionality of data from multidimensional neural recordings of 12 data sets collected in seven different species [141,157]. A small number of PrCs explained most of the variance. PrC1 explained approximately 25% of the variance. Neural activity PrC1 correlated with odorant structure PrC1 and odorant perception PrC1.

The authors emphasized that a principal components analysis is not the only way to represent the multidimensional nature of odorant structure. They identified, for example, a single variable "molecular complexity" as a predictive factor for the perceptual complexity of an odorant and specific neural responses. However, in contrast with the high-dimensional combinatorial code underlying olfaction at the lower stations of the olfactory system, these results are consistent with a simpler olfactory code for the neural activity related to odor perception:

> *"...this finding is consistent with the view of how the mammalian brain*

treats odor: ***synthesizing***[254] *a singular odor percept for an odorant mixture rather than analytically extracting individual odorant features reflecting mixture components".*

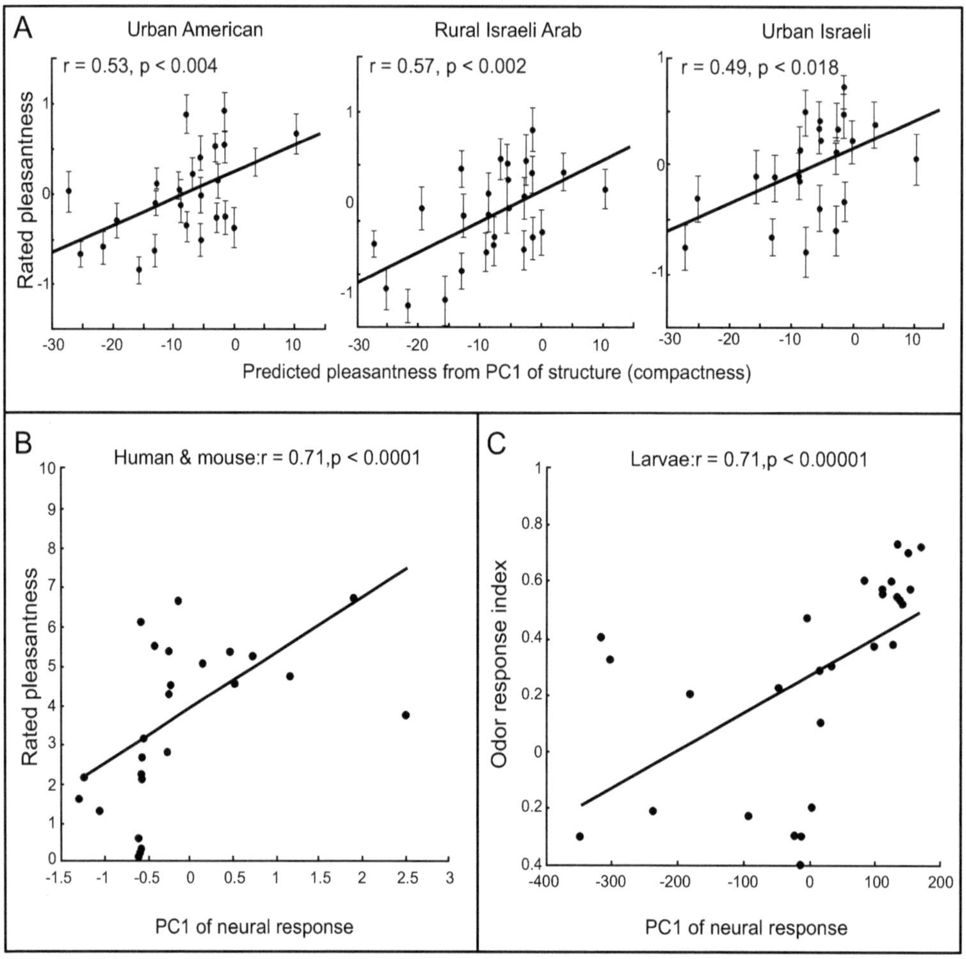

Fig. (8). A metric for olfaction: each dot represents an odor. The metric predicts odor pleasantness across cultures (a). Principal component representing pleasantness is related to PrC of neural response (b). Odor response as behavioral approach or withdrawal is related to PrC of neural response in *Drosophila larvae*. Reproduced with permission of Springer Nature [141].

This line of work indicates that "synthetized" odor percepts, such as pleasantness, have a top-down effect on olfactory epithelium receptor subtypes, thus establishing a connection between perception construct and olfactory epithelium topography [157, 158] (Fig. **9**). As previously discussed, Maeno [123] saw no adaptive advantage for phenomenal consciousness in the absence of episodic memory. Nonetheless, the synthesis of olfactory qualia serves to convey organism-centric adaptive information. As previously discussed, the synthesis of low dimensional percepts supports the transition from the Bayesian calculations of Predictive Coding to phenomenal heuristics. New relations form more compact coalescences [45].

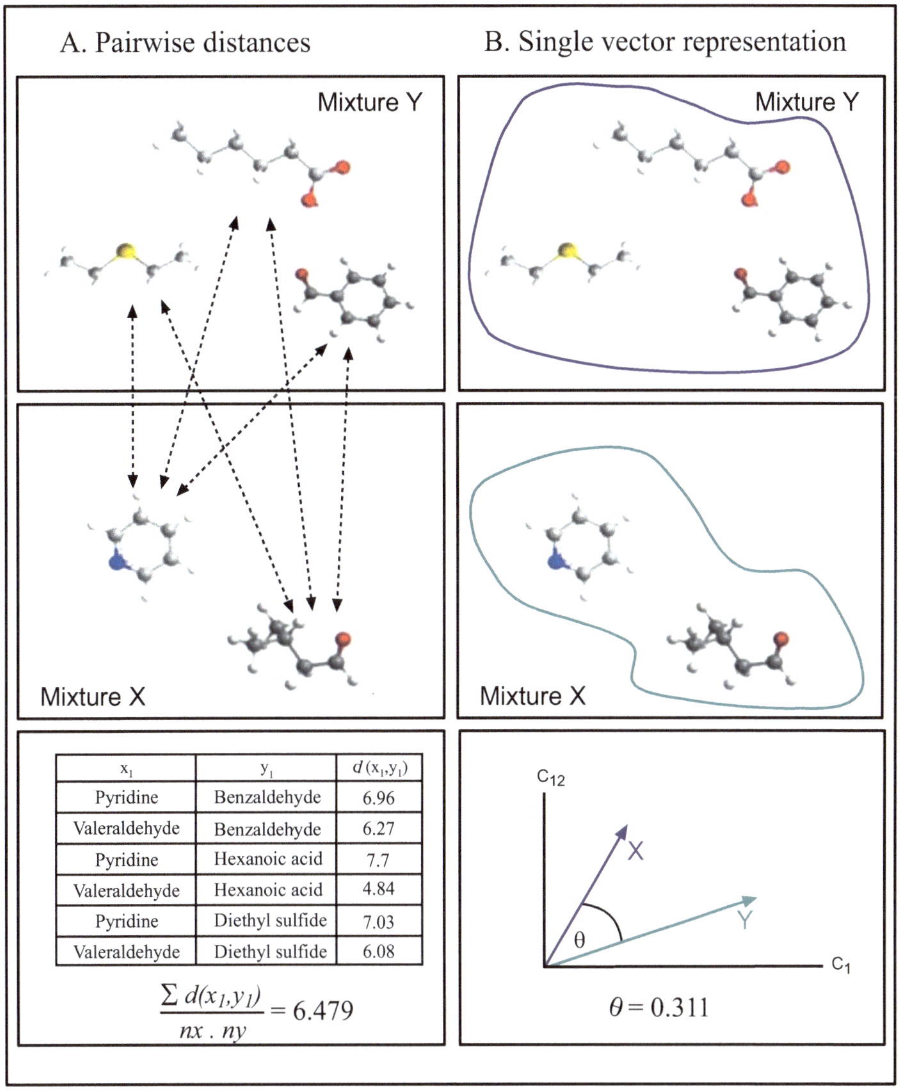

Fig. (9). The synthesis of smell: modeling odorant mixtures as singular objects rather than component amalgamation. Perceptual similarity was predicted only by mixture Y and X being represented by vectors reflecting the sum of their components. Perceptual similarity was not predicted by calculation of the mean of the pairwise distances of all components of the mixtures. Reproduced with permission of Springer Nature [141].

CONCLUSIONS

The transition of external information into organism-centric encoding is made possible by the transformation of information implicit to structural patterns into odor-objects of adaptive value. These odor-objects are crucially dependent on sustained oscillations of distributed and iterative neuronal ensembles. The sparse distribution of odor-induced activity in the piriform cortex resembles the hippocampus. The influence of feedback fibers from higher stations supports the rules of predictive coding. This is particularly important in the PPC, which is heavily affected by these fibers. The PPC seems to be associated with more abstract categories; and odor-objects with experiences of olfactory focal awareness.

The olfactory cortex generates a pattern recognition network out of OB information beyond the structural odorant characteristics. This pattern recognition is influenced by the past history of activity in distributed cortical ensembles[255].

Much like the pattern observed in the hippocampus, PC oscillatory activity establishes cellular synchronicity throughout the ensemble, and allows for the enhancement of connections based on oscillatory phase. Time-dependent plasticity of iterative recurrent networks likely plays an important role in storing and retrieving multiple input patterns and underwriting the binding of spike activity/information produced by mitral and tufted cells coactivation of pyramidal neurons as described by Nagayama and colleagues [132].

Principal component analyses of odor character profiles, odorant structure, and olfactory neural activity identified a simpler code than the high combinatorial code found in the lower stations of the olfactory system. These findings suggest that olfactory perception abstracts regularities from the environment and support qualia as conveying information of adaptive value [159, 160]. This is consistent with the discussion of qualia as providing actual adaptive information and the use of a variable "Q" for its representation. The synthesis of low dimensional percepts supports the transition from the Bayesian calculations of Predictive Coding to phenomenal heuristics.

The olfactory system abstracts structure-based categories from internally generated spatiotemporal maps of the molecular features of odorants. Structure-based categories are further abstracted in odor-object categories [130]. Odor-objects match conscious olfactory experiences. They have a relational origin consistent with this blueprint. Recurrent neuronal ensembles that create gamma oscillations make sustained odor-object images possible. Such a pattern is consistent with phenomenal perception requiring sustained expectations from recurrent neuronal ensembles. Olfactory re-representation in hierarchical abstraction resembles interoception in the insula (Fig. **10**).

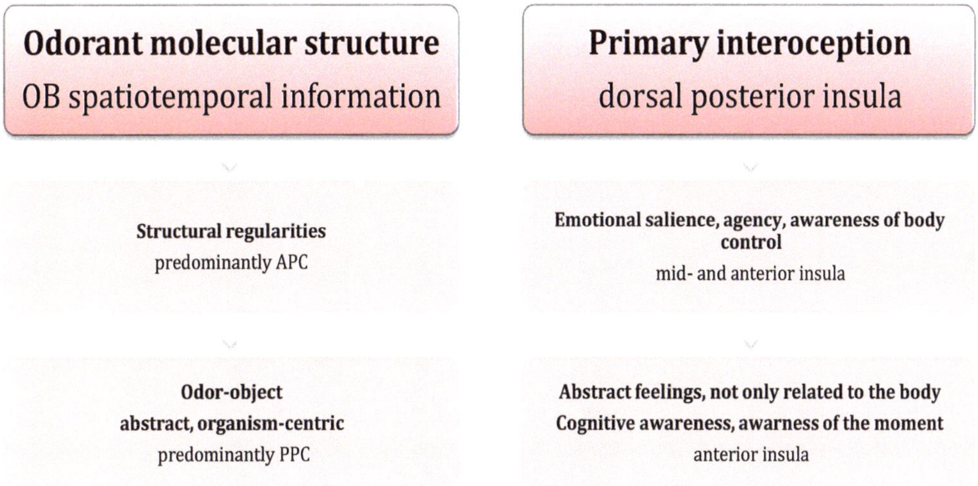

Fig. (10). Parallels between olfactory re-representations and interoception abstraction hierarchies.

A number of computational theories explain how the different parts of the human brain come together to produce human consciousness [46, 47, 61]. A set of common general themes forms the basis of these theories (Fig. **11**). I have previously focused on the IIT and will now analyze how this blueprint fits other theories.

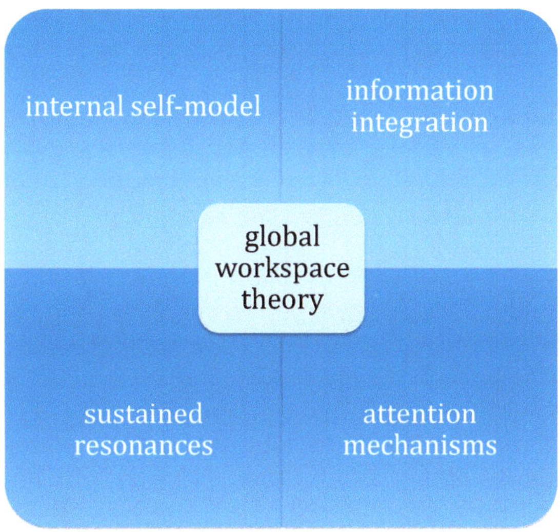

Fig. (11). Computational models of human consciousness.

NOTES

[209] Relating to stimuli outside the organism.

[210] This "off-line" period can be seen as a "sensing phase" or an "expression phase", focused on processing stored information and engaging behavioral responses.

[211] As reviewed by Mori and Manabe, exploratory behaviors are associated with nested LFP gamma-oscillations superimposed on the sniff-paced oscillations of the theta-frequency range (6-10Hz) in the olfactory bulb [78,82]. The relationship between gamma and theta oscillations seems to parallel hippocampal local field theta oscillations occurring during movement, and discussed previously in the section of awareness of location.

[212] And at the vomeronasal organ in the base of the nasal septum or the roof of the mouth in many amphibians, reptiles, and mammals. In many non-human animals, the vomeronasal organ (VNO) plays an important role in the chemical communication between individuals underlying social and sexual functions.

[213] There are several families of odor receptors, such as the vertebrate OR family, which comprises thousands of proteins and part of the G protein-coupled receptor (GPCR) superfamily. In addition, trace amine associated receptors (TAAR) are found in a subgroup of olfactory neurons located in the olfactory epithelium. These receptors recognize odorants with amino groups. They probably belong to the OR family, and signal the presence of predators or dead animals.

[214] The structure arrangement of glomeruli in the olfactory bulb resembles the organization of the neocortex in functional columns [132].

[215] Divided into external tufted cells, middle tufted cells, and internal tufted cells, which display progressively larger bodies as they are located deeper in the external plexiform layer. Mitral cells are located even deeper than the external plexiform layer, at the mitral cell layer. They present with the largest bodies. Tufted cell bodies are located in the superficial and deep plexiform layer. Mitral cell bodies are located in the mitral cell layer. These layers end in the internal plexiform layer and in the granule cell layer.

[216] At mid-inhalation.

[217] The mouse olfactory bulb holds dual representations of each individual OR, typically in glomeruli located at two different sites in the rostrodorsolateral and caudoventromedial regions of the OB. The OB has two "mirror-symmetrical" glomerular maps. Collaterals of Tuft cell axons precisely and reciprocally connect each pair of glomeruli.

[218] A larger synchronization of LFP gamma oscillation is achieved by the synchronized spike discharge of projection neurons and the long lateral dendrites of mitral cells in connection with other activated glomerular modules. In this case, dendrodendritic reciprocal connections between projection neurons and granule cells. In the mouse, about 20 sisters mitral cells associated to a single parental glomerulus form many dendrodendritic connections with mitral cell-targeting granular cells, creating a mitral cell circuit. Mitral cells activation leads to the excitatory activation of granular cells, which synchronously inhibit sister mitral cells, leading to a decrease of mitral cells' activation of granular cells. Mitral cells synchronously fire again either by post-inhibition rebound or by ongoing odor stimulation, generating the cycle underlying gamma oscillation [132]. In contrast, tufted cells project short dendrites in the superficial external plexiform layer to sister tufted cells within the glomeruli. Tufted cells have similar dendrodendritic

connections with dedicated granular cells.
[219] The piriform cortex (PC) is considered the olfactory primary cortex.
[220] Early feedforward and feedback connections in olfaction:
1. Early onset feedforward specificity-projection: tufted cells specificity-projecting circuits convey information with fast gamma synchronization.
2. Recurrent associations in the piriform cortex: recurrent associative PC pyramidal cells store and retrieve olfactory images and provide a mechanism for feedback binding of information.
3. Feedback and feedforward coupling: large-scale feedforward binding by mitral cell signaling and large-scale feedback binding by recurrent associations work together for odor images of objects.
[221] From Nagayama and colleagues [132]: *"As described, sniff-paced early onset and later-onset slow gamma oscillatory activity in the anterior piriform cortex (APC) correspond in timing and frequency with the early-onset fast and later-onset slow gamma oscillatory activity of the olfactory bulb. This observation raises the possibility that the activity of APC pyramidal cells during early-onset gamma oscillation may be the result of the tufted cell axon-Ib association fiber inputs whereas that during later-onset slow gamma oscillation is induced by the combination of preceding tufted cell axon-Ib associational axon inputs and later-onset synchronized inputs from many mitral cells. If so, the later-onset synchronized inputs from mitral cells may cause profound synchronization of those pyramidal cells that had been depolarized by preceding early-onset inputs from tufted cell axon-Ib association axon inputs but have little influence on those pyramidal cells that had not been depolarized by the tufted cell axon-Ib association axon inputs".*
[222] I am alluding to the analogy between realization of information by contrast of different signal mapping by hippocampal place cells and assemblies with the realization of information generated by more localized tufted cell gamma synchronization in the context of larger activation of mitral cell circuitry gamma synchronization. In fact, Nagayama and colleagues [132] emphasize that a similar time lag between fast and slow gamma inputs is also observed in the hippocampus for realization of allocentric information. The fast gamma oscillations of the CA1 hippocampal region receive direct information and synchronization from the medial entorhinal cortex (typically at the trough of theta oscillations). The entorhinal cortex provides information about the animal position as reviewed during the discussion about grid cells. Slow oscillations of the CA3 hippocampal region synchronize with the slow oscillations of CA1, which are maximal at the falling phase of theta oscillations. CA3 is a region credited with storage of positional information. Nagayama and colleagues [132] pointed out that slow and fast gamma oscillations can coexist in the same theta cycle, although they typically occur at different theta cycles. Of note, phase information may be used to discriminate slow and fast oscillations at the hippocampus and the piriform cortex. The difference is that place cells refer to identifying a location in externally existing allocentric space. The olfactory cortex creates olfactory categories based on structural similarities among odorants.
[223] *"Rather than being a futile epiphenomenon of network connectivity, precise synchronization of neuronal discharges seems to be functionally relevant for the binding of distributed responses in a wide variety of neural systems ...why selection based on*

temporal binding would lead to awareness in some cases but not in others. Thus, for instance, both the attentive search for a particular object as well as the visuomotor coordination in a frequently practiced task like driving require context-dependent selection that, according to our hypothesis, may be implemented by synchronization of appropriate neural assemblies. However, although in the former case the selection process usually leads to awareness, this does not necessarily hold for the later. What makes the difference, and what would be a sufficient condition for the instigation of awareness" [137].

[224] The APC is further divided into ventral (APCv) and dorsal (APCd) subdivisions. According to Mori, the major excitatory, afferent flow occurs from the olfactory bulb (OB) → anterior olfactory nucleus (AON) → APCv → APCd → PPC [133].

[225] The olfactory cortex, by definition, comprises all those areas that receive the output of the mitral/tufted cells of the olfactory tubercle in vertebrates. These regions send feedback projections to the olfactory bulb and share information through strong anterior commissure projections.

[226] All olfactory cortices are paleocortical, with exception of the entorhinal cortex.

[227] In a brief summary, as reviewed by Wilson & Sullivan [139]:

Layer 1: plexiform layer - it includes pyramidal cell apical dendrites, mitral/tufted cells axons and accessory fibers, cells of small populations of inhibitory neurons.

Layer 2: cell body layer – superficial pyramidal cell bodies with apical dendrites projecting to layer 1 and basal dendrites into layer 3. (Layer 2a is a superficial component of layer 2 with pyramidal semilunar cells. These cells lack basal dendrites and receive preferentially input from mitral/tufted cells and lesser from association fibers, and do not project back to the olfactory cortex).

Layer 3: cell bodies of deep pyramidal cells, pyramidal cells basal dendrites, interneurons.

Deep to layer 3: Endopiriform cortex (at times considered layer 4) – highly interconnected with the cortex, with dense local and projection connections, with relatively low level of GABAergic interneurons.

[228] *"Bottom-up afferent pathways from the OB are typically associated with top-down feedback axonal projections forming neuronal loops, including OB → AON → OB loops and OB → APC → OB loops ...and top-down inputs from the PC → AON"* [133].

[229] The activity of OB axons-1b decreases during slow wave sleep, an offline period, and deep association circuits generate synchronized spike discharges, associated with sharp waves. Much like the hippocampus, the PC contains a "*combinatorial auto-associative array capable of content addressable memory*" [133]. The PC learns input patterns through spike timing-dependent plasticity.

[230] It is not my purpose to present an expansive review of the OFC (and medial prefrontal cortex) connections, but just to point out that layer 2 pyramidal APC cells project to the layer 1 of OFC, while the OFC projects to the medial prefrontal cortex (mPFC) among other regions and send feedback axons to the APC [133]. The APCd has reciprocal connections with the ventral agranular insular cortex (AIv) and the PPC with the AIv and the posterior agranular insular cortex (AIp), creating loops PC → OFC → PC and PC → AI → PC.

[231] The similarity between the hippocampus and the PC *"implies that the PC might also*

be characterized as a giant cortical module ...The recurrent collateral axons of pyramidal cells in the piriform cortex form excitatory synaptic connections on the dendrites of other pyramidal cells that are distributed widely in the piriform cortex." Feedforward connections tend to terminate in layer 1b (1b association axons) and feedback connections in layers 2 and 3 (deep association axons).

[232] Feedforward connections (OB axon-1b association fibers) become most active during inhalation, including during the inhalation-exhalation transition period. This activation characterizes the early-onset fast gamma oscillation phase mediated by tufted cell circuits and the later-onset slow gamma oscillation phase mediated by mitral cell circuits [131, 132]. The overlap between tufted-1b activation and mitral-1b activation coincides with the inhalation-exhalation transition period. Deep association connections end in layer 2 and 3 and within the deep piriform nucleus. They become most active during the late exhalation period, which is offline.

[233] Food related odors seem to be perceived retronasally. There are about 400 functional OR genes and about 600 (non-functional) pseudo-genes. There are a few trace amine-associated receptors (TAARs) genes as well. Humans have about 5,500 glomeruli in each olfactory bulb with a convergence ratio of more than 10:1, considering the 400 functioning OR genes [141].

[234] The primary olfactory cortex is defined as receiving direct innervation from mitral and tufted cells. It includes the anterior olfactory nucleus, the ventral tenia tecta, the anterior hippocampal continuation of the indusium griseum, the olfactory tubercle, the piriform cortex, the anterior cortical nucleus of the amygdala, the periamygdaloid complex, and the rostral entorhinal cortex, as reviewed by Weiss and colleagues [141]. According to these authors, a "stricter" definition restricts the primary olfactory cortex to the piriform cortex. The piriform cortex is a three-layered allocortex lying (in humans) "*along the olfactory tract at the junction of the frontal and temporal lobes, and continues on to the dorsomedial aspect of the temporal lobe*".

[235] PET: Positron Emission Tomography. Qureshy and colleagues conducted a PET study comparing the functional anatomy of olfactory and visual naming. The OFC activation pattern during visual naming provided evidence for this area's multimodal role [143]. These authors pointed out that the OFC receives information from the visual, gustatory, and olfactory systems.

[236] The authors indicated that the entorhinal neurons of rats display odor selectivity, while the hippocampus in these animals specifically promotes higher-level processing of olfactory information by activating memory traces stored in cortical networks.

[237] BOLD (blood oxygen level dependent): an imaging parameter that indirectly indicates neuronal activity.

[238] Other regions involved in odor identification included the parahippocampus gyrus, the fusiform, the primary visual, and auditory cortices, as well as the inferior temporal and inferior frontal gyri. These findings underscore the importance of different interactive systems involved in odor recognition: a perceptual, a memory episodic, and a semantic system. The fusiform gyrus is under top-down control of the inferior temporal gyrus. The inferior temporal gyrus, according to Kjelvik and colleagues review, guides the retrieval and selection of correct representations [144].

[239] Mizobuchi and colleagues described the case of a patient who suffered from complex

partial seizures [145]. This patient experienced unidirectional olfactory aura as a result of an ipsilateral unruptured brain aneurism (an enlarged artery due to weakened walls) that exerted pressure on the right OFC. The patient smelled sweet pleasant odors, such as those that emanate from sweet fruit or flowers, during the aura that anteceded epileptic seizures. The smell always seemed to come from her right side. MRI and CT angiography (MRI and CT angiography: Magnetic Resonance Imaging and Computed Tomography that visualize arteries and veins) identified the aneurism at the bifurcation of the right middle cerebral artery, and the authors intrasurgically recorded small EEG spikes in the OFC, the superior temporal gyrus and the uncus. Li and colleagues reported the case of a patient who sustained a traumatic injury to his right OFC [146]. The patient had complete anosmia and no awareness of odors presented to his right and left nostrils. However, MRI imaging indicated that the left OFC responded more intensely to stimuli presented to the left nostril in comparison with the right nostril. These findings agree with the concept of "blind smell", a measurable response to smell in a person who cannot consciously sense odors. The authors also observed a modest odor-related activity in the PCs, suggesting that the trauma did not affect the PC circuitry. A lack of attention to smells due to right OFC damage supposedly caused the modest activation in response to odors. These two case reports suggest that the right OFC provides the local circuitry underlying conscious olfactory experiences. From the perspective of the IIT, the right OFC can be interpreted as encoding information regarding the conscious experience of odors. Damage to the right OFC stops the brain from globally integrating this information. Nonetheless, olfactory deprivation has different effects on the PC and the OFC.

[240] Wu and colleagues assessed the consequences of seven-day olfactory deprivation in humans [147]. They found that deprivation causes reversible changes in odor-evoked fMRI activity both in the PPC and the OFC. However, behavioral olfactory performance remained stable despite the substantial decrease in odorant afferents. Odor deprivation had no impact on the mean fMRI signal change in the APC, but reduced the odor-evoked activity in the PPC and increased activity in the OFC. Deprivation selectively impaired odor quality coding in the OFC, but not in the PPC. According to the authors, "*given the reported role of human OFC in olfactory attention, it is also possible that the interruption of odor input (as indexed by PPC activity) places greater demands on OFC to focus attentional resources on the incoming stimuli* ".

[241] The OFC has a strong connection to the PC. The transformation of odorant information from the olfactory bulb into structure-based odorant categories (APC) and into "odor-object" categories (PPC) occurs under feedback influence of the OFC. Conversely, epileptic discharges stimulate the OFC. In other words, realization of encoded information involves both the OFC and the PCs, much like observed in the thalamic-cortical unit of the dynamic global workspace (dGW).

[242] According Wilson & Sullivan [139]: "*the afferent and intrinsic synaptic inputs result in sparse, spatially distributed pyramidal cell odor-evoked activity*". The role of activity dependent plasticity renders the development of reliable ensembles to depict familiar odors, activation of the whole ensemble with limited or degraded information (cortical pattern completion), which promotes perceptual stability and, conceivably, stability of qualia. Much like the pattern observed in the hippocampus, PC oscillatory activity establishes cellular synchronicity throughout the ensemble, and allows for the

enhancement of connections based on oscillatory phase. Time-dependent plasticity of iterative recurrent networks likely plays an important role in storing and retrieving multiple input patterns and underwriting the binding of spike activity/information produced by mitral- and tufted-cell induced activation of pyramidal neurons [133,159].

[243] As reviewed by Wilson and Sullivan [139]: the initial phases of processing OB inputs may happen at the anterior olfactory nucleus (AON), leaving for the PC later stages of building odor-objects.

[244] Unlike other sensory systems, the olfactory system bypasses the obligatory thalamic relay connecting sensory receptors and cortex [148].

[245] According to Howard and colleagues [150] results, the spatial arrangement of fMRI activity in human PPC is distributed and overlapping without any obvious clustering, suggesting that perceptual information about odor quality is reflected in ensemble activity within the PPC. The authors conclude that olfactory codes for odor-objects are arranged similarly to visual-object categories ("houses, cows, chairs") in the inferotemporal cortex.

[246] The olfactory cortex enables the conscious experience of olfaction when it reads the molecular feature maps represented at the OB, including the temporal coding of odor quality and intensity [151]. Second-order projection neurons transfer odorant information represented at glomerular maps to higher olfactory centers in the forebrain [139]. Higher olfactory centers receive the information in specific manners: nonselective or biased, diffuse or convergent [151].

[247] However, some olfactory information is the product of more specific receptor stimulation, such as with the trace amine-associated receptors (TAARs), which are expressed at a subset of olfactory sensory neurons (OSN). TAARs are a separate class of OR. They are sensitive to odorants with amino groups, which signal the presence of dead animals or predators. Among other regions, they are found in the Grueneberg ganglion (GG) located in the entrance of mammals noses as a subset olfactory system dedicated in detection of alarm pheromones and cold temperature [132].

[248] Although feedback control is already present at lower levels, such as on cortical feedback projections on olfactory bulb circuits [152].

[249] Phenomenal objects provide flexible mnemonic devices supporting long-term representations storage [56].

[250] With the exception of the olfactory tubercle, the olfactory cortex (OC) has a pyramidal cell organization. Olfactory bulb projections connect with superficial dendrites of layer 2 and 3 at layer 1 as feedforward information. Projections of association fibers connect at layer 1b as feedback information. Pyramidal efferents at layer 2 and 3 at the anterior olfactory nucleus and PC "massively" project to the olfactory bulb as feedback signaling to the olfactory bulb [130]. Layer 2 pyramidal cells at the anterior PC project feedforward information to the ventrolateral orbital cortex, the lateral orbital cortex, and the ventral agranular cortex of the insula. The OFC projects feedback axons to layer 3 of the PC. I am not including here detailed information about indirect thalamic pathways, the lateral entorhinal cortex, the amygdala, or the ventral pallidum. It is still unclear how these complex large-scale networks transition from inhalation to exhalation, although the basic architecture is not incompatible with predictive coding. However, while the primary posterior insular afferents are already "*fundamentally about the body*" [105,107], olfactory afferent information is not implicitly embodied but modeled by olfactory

activity-dependent refinement during development and adulthood [133].

[251] Granted potential exceptions, such as information conveyed by trace amine associated receptors (TAAR).

[252] Although the origin of qualia should be traced to information for adaptive behavior, not all qualia should be considered adaptive information. Stephen Gould [307] discussion of the arches and spandrels of evolution, comparing secondary products of adaptation (not necessarily adaptive) to masonry spandrels applies here.

[253] As described by Weiss and colleagues [141]: the PCA applies data from N-points in a M-dimensional space (138 odorants in 146-dimensional language odor descriptors) in new dimensional spaces, called principal spaces that successively explain the maximal portion of variance.

[254] The emphasis is from Weiss and colleagues [141].

[255] According Wilson & Sullivan [139]: "*the afferent and intrinsic synaptic inputs result in sparse, spatially distributed pyramidal cell odor-evoked activity*". The role of activity dependent plasticity renders the development of reliable ensembles to depict familiar odors, activation of the whole ensemble with limited or degraded information (cortical pattern completion), which promotes perceptual stability and, conceivably, stability of qualia.

CHAPTER 16

Attention, Neuronal Oscillations and Predictive Coding

Abstract: Theories of consciousness share the idea of consciousness realization in neuronal states. Conscious experiences take place in embodied agents that generate an internal model of the external environment and of themselves. Artificial intelligence researchers have focused their efforts on the creation of a multi-level consciousness-based architecture that follows these precepts. Attention mechanisms can serve as control models on silicon environments with limited resources for information processing. Conversely, attention can emerge when information-processing modules cooperate to solve a given problem. However, computationally responding to a topic does not equal conscious processing. Top-down attention and consciousness are conflated, but not the same. Phenomenal consciousness seems possible with activation of local networks, but it is fully realized only after global ignition. Conscious experiences have their own abstraction level. However, they depend on the dynamic cross coupling of neuronal oscillations across processing levels. Predictive Coding ties efferent copies of corollary discharges to the experience of agency. A system based on monitoring action-based signals may underlie both ownership and awareness. An efferent copy is an extra copy of a given command from a control region. It informs local processors of intentions, allowing these downstream processors to timely match intention with the actual results of motor commands. This voluntary motor model applies to abstract experiences. A generative model depends on actions. It maps hidden causes to sensory consequences of these actions. Perception corresponds to the inverse mapping from sensations to hidden causes. A subject may be inferred from inverse mapping of the hidden causes that underlie seen or unseen globally integrated actions. The subject represents the narrative or scenario that predicts sequences of events. However, caution should be applied against strict localizationism in the interpretation of this model.

Keywords: Access Consciousness, Adaptive Resonance Theory, Artificial Intelligence, Attentional Models, Being the Subject as Inference, Bottom-Up Awareness, Consciousness-Based Architecture, Corollary Discharge, Cross-Frequency Coupling, Emotional Tagging, Feelings of Knowing, Generative Model, Global Workspace, Labeled-Line Limitations, Localizationism, Machine Consciousness, Neuronal Oscillations Summary, Phenomenal Consciousness, Primary Visual Cortex, Predictive Coding, Self-Modeling, Systems 1 and 2, Top-Down Amplification, Variable Q.

Paulo J. Negro

ATTENTIONAL MODELS

Humans consciously detect only a portion of the available information about their bodies and the external world. Mechanisms for the active selection of relevant information provide models for conscious processing of information. Attention has a high correlation with conscious awareness [61]. Computationally, attention mechanisms can select representations in topographically structured networks.

Top-down task-dependent control of attention plays an important role in human consciousness[256]. However, attention does not necessarily trigger phenomenal experiences and attention without consciousness was found in top-down and bottom-up, spatial, temporal, feature-based, object-based and eye-based attention paradigms [161].

Attention mechanisms can serve as control modules in silicon environments with limited resources for information processing. Conversely, in Global Neuronal Workspace-like silicon systems, attention emerges when specialized information-processing modules cooperate on a given problem. Models with limited processing resources that mimic the multilayer architecture of the cerebral cortex are considered particularly helpful simulations of this paradigm.

However, this framework equates attention and conscious awareness. In other words, in these computational models, attending to a topic equals conscious processing [61]. The same does not seem to occur in the neuropsychological environment of the human brain: *"awareness treats the items at the attentional level as symbols that can be manipulated"* [40]. Such idea can be considered consistent with the transition into a heuristics-based phenomenal consciousness. Conscious experiences have their own higher abstraction level.

Nonetheless, such models may test the relationship between specific aspects of attention and conscious experiences. For example, represented symbolic concepts tag lower-level representations in the information stream. From a computational perspective, tagging constitutes a powerful tool to understand how the brain processes information [40]. Emotional labels can tag streams of information. Affective tagging may play a crucial role in the recognition of a loved one[257]. When a person thinks about what someone else might be thinking, the generated simulations use networks that support other forms of thinking. An affective tag of "mine" or "other" reduces the risk of confusion with such simulations[258].

As we will see in the section on the Adaptive Resonance Theory (ART), attentive matching is a crucial aspect of matching top-down expectations [162] by means of *"top-down on-center off-surround circuit activation to read how expectations match bottom-up feedforward patterns[259]"*.

Many daily behaviors [161] are conducted without full awareness[260], including automatic behaviors, such as driving a car, which would not be possible without phenomenal awareness. The primary visual cortex is considered necessary, but not sufficient for visual consciousness. However, some stimuli may become conscious without top-down boosting, depending on the context of the scene (visual cluster, contrast, stimulus duration). Local and regional feedback loops sustain this kind of activity, which is less vigorous and of easy decay[261].

The pattern described above clearly differs [161] from a model whereby visual phenomenal consciousness emerges from wide-ranging coalitions, boosted by feedback fibers originated at the anterior regions of the brain and made fully accessible to consciousness. These coalitions become part of a reciprocal reverberatory neuronal activity that lasts longer than the initial stimulation and involves the prefrontal cortex and other regions that run work memory and planning.

I pointed out that the realization of information into forms of awareness represents a relationship between expectation and information in the context of the strange loop *'expectation.information'expectation.information*. Such strange loop underlies the realization of local information into the variable Q or consciousness-information, later realized in the overall consciousness context. Overall, the idea is consistent with local realization having limited access to consciousness as well as with its amplification towards sustained and global reverberatory states. The origin and hierarchical nature of feedback fibers is of central importance to hypothesis generation in self-modeling architectures.

INTERNAL MODELING

Computationally, subjective conscious experiences arise from agents that have an internal model of the external environment and of themselves. Conscious experiences can occur in embodied agents capable of these self-models. As reviewed by Reggia, this consideration has a special relevance for the study of robots controlled by recurrent neural networks [61]. Such robots can handle sequences of stimuli targeting predictions of the future based on past and present information. In fact, an "expectation-driven" robot seemed to recognize itself in a mirror[262]. Whether true or not, it was claimed that the robot had instantiated consciousness.

FIRST ORDER AND HIGHER ORDER REPRESENTATIONISM

The concept of self-modeling is crucial, but not enough for the understanding of human consciousness. It has been postulated that conscious experiences use a higher level of representation[263] than unconscious mental activity. A change of representation level may also hold true for experiences of awareness transitioning from Kahneman's [118, 119] rapid "System 1" to the deliberate "System 2".

First-order representationism considers any conscious state a representation, and a conscious experience the product of this representation's content [163]. From this angle, human consciousness directly results from first-order representations distributed in a global space of activity. However, according to higher-order representationism, higher order representations of these first-order perceptual representations determine the contents of consciousness[264].

Based on this understanding, artificial intelligence researchers have focused their efforts on the creation of multi-level "consciousness-based architecture" [61]. Such architecture underlies the perceptual processing of information and provides a scaffold for metacognitive networks that represent lower orders as primitives for manipulation. However, neither first-order nor higher-order representationism may provide a complete model of consciousness. Computers perform hierarchical and relational processing of information, including symbolic processing. Nonetheless, most people and most researchers do not consider computers conscious despite their processing of information [61].

A given computational perspective does not necessarily exclude the others. A model that acknowledges a role for information integration, self-modeling, hierarchical representations, and a global workspace, solidly anchored in the neurobiology of the human brain, more comprehensively captures the nature of consciousness than any of these models individually. Monitoring mechanisms do not necessarily promote consciousness, but they may tag computation mechanisms associated to consciousness at the appropriate level of abstraction.

COMPUTATIONAL MODELS AND PREDICTIVE CODING

Predictive Coding, when applied to brain macrodynamics, ties efferent copies, or corollary discharges, to the experience of agency. A system based on monitoring action-based signals may conflate ownership and awareness [61].

Computational models of feedback and feedforward processing of information consistent with the principles of Predictive Coding have already successfully applied control theory to robotics [61]. An "efferent" copy represents an extra copy of a given command from a control region. The control region sends the efferent copy downstream, informing input regions of its intention. The efferent copy provides information to local processors that timely match intention[265] with the actual results of motor commands.

As discussed in the Predictive Coding section, evidence supports the existence of efferent copies, or corollary discharges, in the human brain. Corollary discharges are more traditionally applied to motor behavior models. Executive regions in the brain send downstream volleys of actionable motor instructions that generate specific movements[266]. Proprioceptive and visual systems inform about the outcome of these instructions. The executive regions also send out an efferent copy of the motor instructions, the corollary discharge, to serve as a forward basis for monitoring the predicted movement. Feedforward monitoring acts faster at lower levels of information processing.

This voluntary motor model also applies to abstract experiences and behaviors. The paradigm, applied to the Global Neuronal Workspace, sees that frontal executive regions send corollary discharges to the cingulate and parietal cortices, with an inverse model of the executive instructions.

BETWEEN THE EASY AND THE HARD PROBLEMS

A distinction between "easy" and "hard" problems of consciousness has yet to be clearly established. As described in the beginning of this book, easier problems include tasks such as processing information, focusing attention, controlling mental states, whereas the hard problem is the explanation of the origin of subjectivity, itself. However, local processing of phenomenal objects includes tasks related to the "easy" problem of consciousness.

Subjective experiences may characterize abstract forms of awareness that do not have an association with qualia, in a strict sense. For example, awareness of location, feelings of knowing (FOK) and other abstract forms of awareness described in the Global Neuronal Workspace theory fall into this category. These types of awareness can be seen as resulting from computational mechanisms, but do not necessarily explain the origin of qualia. They sit between the "hard" and "easy" problems of consciousness.

A BRIEF SUMMARY OF NEURONAL OSCILLATIONS

Phenomenal objects constitute an organized hierarchy of patterns of self-presenting phenomenal features and take part in the dynamics of a phenomenal world characterized by a phenomenal space that centers on the subject[267]. This phenomenal space or 3D matrix arises from a unifying 3D coordinate system bridging nonconscious biological mechanisms and phenomenal consciousness [54].

Consciousness emerges from a hierarchy of operations, ranging from blind neuronal operations, to elemental functional assemblies that support cognitive operations as well as complex macro-level operations that support integrated phenomenal experiences. An operation represents, at the same time, an act (or process) and an object (or thing). Conscious experiences are "transparent" representations through which the subject apprehends the characteristics of the object without noticing the subsumed nonconscious mechanisms that underlie the phenomenal experience.

Widespread phase synchronization of gamma oscillations correlates with conscious perception [164]. Perceptual switching during binocular rivalry is time-locked to gamma-band synchronizations and recurs at a theta rate. The onset of new conscious percepts coincides with the emergence of new gamma-synchronous assemblies locked to an ongoing theta rhythm [165]. Furthermore, widespread gamma oscillation can originate from the thalamus, as it will be discussed later [166].

Coupling of oscillatory activity underlies the selection of signals that provide an access to consciousness, structure conscious contents, and create a global workspace for consciousness to emerge [150, 167 - 171] (Table **4**).

Table 4. Simplified summary of oscillatory activity [172, 173].

FREQUENCY BANDS	PERCEPTUAL, SENSORIMOTOR, COGNITIVE OPERATIONS
Delta (0.5-3.5Hz)	Motivational processes, reward system, deep sleep phases
Theta (4-7Hz)	Working memory, emotional arousal, fear learning, attentional sampling
Alpha (8-12Hz)	Relaxed conscious state with eyes closed: awake resting-state without sensory inputs, disengagement task irrelevant areas, short term memory retention
Beta (13-30Hz)	Alert conscious state with eyes open: sensorimotor processing, working memory, top-down actions
Gamma (>30Hz)	High peak cognitive performance: perceptual integration and binding, stimulus selection, movement preparation, memory formation, attention, consciousness

As reviewed by Engel and Fries [172], consciousness seems to require large-scale reverberations involving frontoparietal regions that implement top-down actions. These oscillations integrate lower and higher brain stations as well as top-down and bottom-up influences.

Neuronal oscillations have an association with levels of consciousness, with the selection of sensory information that provide an access to consciousness, as well as with the contents of consciousness. Beta band oscillations are associated with attention selection and other conditions involved in strong top-down processing of information[268]. The association includes long-range beta band coherence between the frontal and parietal cortices.

Gamma band[269] synchronizes local and long-range functional communications [174]. Attention enhances gamma band phase coherence in regions responsible for the attended stimuli while reducing coherence in the alpha and beta bands. Gamma synchrony is widely reported as associated with conscious detection of stimuli [175, 176]. Large-scale gamma-band synchrony may be the oscillatory substrate for the stream of consciousness [165, 175, 176].

Fontolan and colleagues analyzed intracortical activity with macroelectrodes implanted in the primary auditory cortex, and in the association auditory cortex in three patients with epilepsy[270] [126]. They found that information travels top-down (T-D) and bottom-up (B-U) using different frequency channels[271].

Hyafil and colleagues used a model of auditory cortex function to assess whether coupled theta-gamma oscillations could underlie a multi-scale analysis of speech [177]. They found that the gamma network could encode temporal patterns of speech if entrained by the theta rhythm, which, in turn, was driven by syllable boundaries[272].

Theta-gamma coupled[273] oscillations may therefore de-multiplex[274] speech. Gamma spiking neurons within theta bursts encoded the temporal patterns of language. Their simulation supported other empirical studies that showed cross-frequency coupling properties[275]. Instead of space and time, which form the content of information encoded by the hippocampus during navigation, the de-multiplexing of speech takes effect by encoding time-events and time. Oscillations can provide not only the expectations of "what" in Predictive Coding, but also expectations of "when" in predictive timing [178 - 180].

Cross-frequency coupling has the characteristics of a hierarchical organization, where low frequency oscillations may function as instruments for sensory selection [181, 182]. The brain has a bias towards a rhythmic mode of operation[276], which allows sensory cortical entrainment, phase-locking to the temporal structure of attended stimuli as well as the alignment of high-excitability oscillation phases with the attended stimuli and its temporally associated events. Cross-frequency phase-linking represents widespread neocortical signaling, particularly theta-gamma and alpha-gamma couplings [183].

In summary, dynamic coupling of neuronal activity selects, integrates and structures signals conveying the level, the access, and the content of consciousness. Coherence among neuronal assemblies allows the temporal signature of consciousness and the competition among different assemblies for informational access to consciousness. Space-time oscillatory coupling can frame expectations and convey information across computational and noncomputational hierarchies.

INTEGRATION OF LOCAL AND GLOBAL ACTION

Evidence suggests the existence of integration without consciousness [47, 184, 185], but full consciousness requires long-range integration [173]. Therefore, gamma oscillations may or may not convey conscious experiences. It depends on the context[277].

That consciousness should require both a local origin and a globalized access constitutes a paradox yet to be resolved. Different theories place the origin of conscious experiences in sensory areas[278] [162, 163, 186], in post-sensory areas[279] [187], or both.

While sensory processing alone does not generate consciousness, global processing requires information processing by specialized local architectures. Consciousness-information becomes integrated as a conscious experience by the joint action of localized networks and global broadcasting[280] [46, 47, 61, 183, 188, 190, 191, 205, 211-214]. A global workspace seems to require a complementary concept of localized consciousness-information.

Most theories propose a combination of local and global dynamics underlying conscious states. Information becomes conscious after triggering a high-level decision to engage multiple brain centers [46, 47, 189]. However, no central registry reconciles the assembly of information that is broadly encoded in the brain [191].

Consider the following thought-experiment. Imagine that a brain-scale assembly of information to be experienced as a conscious event will connect to a single point. It does not matter if this point physically lies in the brain, in a mathematical dimension, or if it is projected into infinity. All the information converges to this point to be assembled as an integrated whole. Organizing this multitude of informational pieces requires a registry of their origin, and of the immense variety of activation patterns encoding all possible representations in exact timing and speed. Still, there is neither a central registry nor a single point of consciousness.

LOCAL AND GLOBAL INTERACTIONS IN VISUAL AWARENESS

Neuronal activity in the primary cortex and higher areas of the visual system correlates with changes in the contents of visual awareness [175]. There is no clear-cut distinction between areas that support conscious and nonconscious processing [193]. Both conscious and nonconscious visual experiences are associated with the activity of higher functionally specialized visual regions, such as the middle temporal visual area (V5/MT). Nonconscious visual stimuli can affect behavior, such as in the case of words "masked" by experimental design

[200]. Such observation implies that nonconscious visual stimuli can be processed at the point of semantic identification and object identity.

The primary visual cortex (V1) is not considered to be a correlate of visual consciousness. Full conscious access is granted by increased activity in the ventral visual cortex, parietal ignition, gamma synchrony, and increased modulation of parietal-occipital alpha rhythm [175,195-199]. Effects on visual perception overlap with attentional modulation, but attention seems not to be sufficient for visual awareness. Top-down effects are found in experimental paradigms that induce illusions, such as "phantom" and fill-in images. These experiments show both changes in the primary visual cortex and the activity of top-down feedback fibers. Recurrent top-down and bottom-up processing supports visual perception [197, 267, 268].

Primary visual cortex (V1) responses can thus reflect subjective experiences regardless of not being considered a neural correlate of visual consciousness [198]. The primary visual cortex seems necessary, but not sufficient for visual consciousness [193,198-201]. Damage and disruption of frontal and parietal cortices can cause visual extinction and neglect. The act of imagining an image triggers a pattern of activation that resembles the pattern elicited by visual stimulation. Altogether, these considerations support not only a modulatory effect, but causal role for feedback signaling in visual awareness.

REVISION OF CORTICAL SIGNALS OF CONSCIOUSNESS WITH INTRACRANIAL ELECTRODES MEASUREMENTS: A HYBRID LOCAL AND GLOBAL MODEL

Conscious events correlate with transient large-scale long-distance synchronization of gamma oscillations [164]. The synchronization occurs early (~100ms) and may trigger other neurophysiological measures associated with conscious perception. These measures include the enhancement of the P3a-like component of the evoked response potential to a stimulus, and theta oscillations over frontal regions.

According to the Global Neuronal Workspace theory, conscious perception results from global activity. Local cortex activation, a preparatory stage, may or may not trigger the global event. In visual perception, early activation of the visual cortex occurs at ~100ms, followed by the activation of higher-order content-specific visual areas at ~200ms. Finally, the global activity that characterizes conscious perception occurs at ~300ms.

Local models of conscious perception do not require a 300ms late phase in conditions of passive viewing conditions, *i.e.* conditions not requiring the subject to take actions after the stimulus. On the other hand, in global models conscious perception requires a late phase. Although not mandatory, a post-stimulus late phase may occur in local models as the result of non-visual cognitive events that are secondary to perception.

Noy and colleagues studied 43 patients and 4,445 recording sites using large-scale intracranial measurements[281] [203]. In simple visual recognition, they identified short latency intense visual responses over the entire cortex. Responses begun in the visual cortex and continued in higher-order visual regions at ~200ms. No delayed visual responses, at 300ms, could be detected. In contrast, repeated images that require a motor response were associated with a delayed ~300ms peak of brain high frequency (BHF)[282] in the frontoparietal cortex[283].

Unexpectedly, an ultra-fast global signal of lower magnitude spread throughout frontoparietal regions. The signal was not restricted to visual processing regions[284]. The signal occurred at ~200ms, simultaneously with the activation of higher-order content-specific visual areas.

These results do not fit with strict local and global theories of conscious perception. The absence of a global, large-scale BHF response at ~300ms in visual perception provides evidence against the global models of conscious perception. However, a global event at ~200ms is not consistent with theories of localized conscious perception.

The study challenges the hypothesis that global events at ~300ms post-stimulus represent a “signature” of consciousness. The ~300ms scalp EEG finding does not necessarily represent a ~300ms intracranial BHF response. In particular, a late ERP peak and post-perceptual processes associated to task completion could contaminate such scalp-level measurements.

A hybrid model for local and global dynamics may be necessary [198,203]. Conscious access to sensory stimuli correlates with activity in higher associative cortices, including the prefrontal and anterior cingulate regions. Sudden parietofrontal activation marks conscious perception. Anterior activation occurs in parallel with top-down amplification [202] and increase of functional activation of stimulus-specific areas.

Nonetheless, there can be stimuli awareness without such top-down boosting [161]. Local and regional feedback sustains this activity, which is less vigorous and of easy decay. Faint perceptual experiences can be generated by localized feedback, while wide-range coalitions conveying higher-level hypotheses are necessary for sustained conscious experiences [202, 204, 205]. Awareness is thus seen as crucially dependent on the quality of the feedback fibers conveying top-down hypotheses.

"BEING THE SUBJECT" IS THE REALIZATION OF INFERENCE INVERSELY MAPPED OUT AS HIDDEN CAUSES OF GLOBALLY INTEGRATED ACTIONS

Information must be synchronized in time and space to become conscious. Information integration, representation mechanisms, internal modeling, general and specific attention are all valid computational perspectives. However, brain functions are the product of evolutionary constraints over long periods of time. The brain is not *"assembled as a computer with prefabricated parts fitted together in accordance to a blueprint that exactly specifies the nature and purpose of each circuit and switch"* [67]. It is unlikely that computing routines will capture its complexity even if relatively simple simulations show heuristic value.

Oscillatory phenomena are necessary for the sustained brain states capable of sustained states of awareness. Cross-frequency coupling at multiple spatial scales has the characteristics of a hierarchical organization and fits the idea of realization of information. However, the rhythms are likely a single coalescence emerging out of non-computational subcellular phenomena. Mathematical descriptions of oscillatory wave-like phenomena may have heuristic value but should be connected with the realization of information by inference mechanisms across abstraction levels.

According to the principles of Predictive Coding, generative models depend on actions. They map hidden causes to the sensory consequences of these actions. Perception corresponds to the inverse mapping from sensation to the hidden causes of relational abstractions. I propose that this model should be extended to predict that the experience of "being the subject" is the inference inversely mapped out as hidden causes of the global integrated actions of the organism. Keep in mind that nonconscious inferential narratives already underlie Kahneman's "System 1", which is responsible for the fast tracking of human mind appraisals [118, 119, 206].

In this view, the subject is the narrative or scenario that predicts sequences of events. It emerges as the consequence of living organisms' statistical physics. At this high level of abstraction, the hidden causes ϑ selected by the ensemble density $q(\vartheta,\lambda)$ can be understood as the subject, itself. The subject thus exists in emptiness, as inferred from the integrated hidden causes of the organism's visible and hidden global behaviors. Consistently with the *expectation.information* loop, this paradigm depends on potential states.

THE LOCALIZATIONISM ERROR OF MY WAYS

One may approach the paradox of local content generators associated with widespread activation by imagining a subject who actively realizes information from local generators during the experience of a conscious percept. Lessons from corollary discharges and predictive coding theory apply here. In this sense, areas that process target stimuli should indeed display enhanced activity.

Realization of information into conscious perception happens across a hierarchy of abstraction, which, conceivably, takes local processing representations as primitives into the higher-level global processing of the conscious experience. Codes become more and more abstract and relevant to behavioral goals. There is no quale stored in local network representations. Qualia appear *de novo* as these local representations are realized into the global workspace. Realization is the transformation of information across different levels: from structures to functional relationships, from functional relationships to abstract functional relationships, which eventually vary according to each other, rather than to their conditions of origin.

This view is cryptically localizationist. What do I mean by this? The brain is traditionally seen as a combination of diverse functionally specialized areas. For example, the cerebral cortex has subregions specifically involved in different sensory and motor activities, while other areas take part in the processes of understanding, producing speech, as well as visual and auditory perception. The stimulation of the primary motor cortex, anterior to the central sulcus causes muscular contractions, whereas the stimulation of the primary somatosensory cortex, posterior to the central sulcus, gives origin to bodily sensations. The idea of realization of abstractions across a neat computational hierarchy implicitly emphasizes this traditional view of brain functions.

THE LIMITATIONS OF THE LABELED-LINE CONCEPT

However, more precisely, information is best seen as distributed in the brain:

> *"There are no absolute or fixed spatial borders between cortical areas that dictate or constrain the functional operation of the cortex as a role. Instead, the cortex should be embraced as a formidable, but finite, neuronal space-time continuum. In this continuum, functions and behaviors are allocated and produced respectively by recruited chunks of neuronal space-time, according to a series of constraints, among which are the evolutionary history of the species, the layout of the brain determined by genetics and early development, the state of the sensory periphery, the state of the internal brain dynamics, other body constraints, task context, the total amount of energy available to the brain, and the maximum speed of*

neuronal firing" [207].

Despite cortical specialization, evidence derived from brain machine interface (BMI) studies challenge a paradigm based on localizationism [207]. Cortical regions are not purely unimodal. In fact, multimodal processing of information starts in much lower stations. Primary visual cortex is enlisted to process auditory stimulation if vision is disrupted[285]. Both primary visual and primary somatosensory cortices display cross-modal responses. BMI studies measuring the concomitant activity of neuronal arrays show that *"a much smaller sample of primary motor cortex neurons can be used to predict hand position, at much higher levels of accuracy, than equivalently sized population of neurons in a region of the posterior parietal cortex (PPC)."* In other words, information is distributed, and neurons can conceivably multitask[286].

It is not my intention to review the nuances and controversies between localizationism and distributionism. Nonetheless, the idea of neat computational hierarchies anchored on "labeled lines" does not truly describes how the brain works. Most likely, a "universal coding strategy" is shared by the neocortices. Spatiotemporal firing patterns emerge from a *"continuous, salty, and hence highly conductive space ...generating electromagnetic fields"* capable of affecting neighboring neurons [207]. Overall, a stance leaning towards distributionism speaks against a linear hierarchy of abstractions, but towards a confluence of levels of abstraction[287].

The complex multitude of rhythms described as cross-coupling at multiple levels actually represents a single coalescence[288] [174,182, 207-210]. A phase-locked neuronal coalition can concomitantly convey information and frame expectations at cross frequency with another rhythm. These rhythms emerge out of rhythms, and all of them emerge from subcellular phenomena underlying the disruption and restoration of the electrostatic equilibrium of the cellular cytoplasm [41]. The complex wave-like manifestation of these oscillations depends on subcellular processes that are not based on integers. Taken together, these considerations can be seen as supporting an ultimate noncomputational origin for consciousness.

CONCLUSIONS

Progress summons the reconciliation between Predictive Coding, computational models of consciousness, and the oscillatory neuronal framework of human consciousness. Predictive Coding calls for iterations between predictions (or expectations) and predictive-error (or information). Signals conveyed by dynamic oscillatory coupling thus enact these iterations. Loops described as *'expectation.information'expectation.information* fit a paradigm that allows top-down and bottom-up connection as well as the realization of information across abstraction levels. In this context, dynamic oscillatory coupling can simultaneous signal expectation and information across computational and noncomputational levels. Sustained neural assemblies' oscillations reflecting phenomenal objects can make the transition towards logic dominated by heuristics. Global ignition thus activates higher-level feedback fibers that enhance assemblies building up those phenomenal objects made conscious through sustained resonances.

NOTES

[256] And synchronization is subject to attentional influences.

[257] One of the explanations of certain psychiatric misidentification syndromes, such as Capgra's syndrome in which the patient develops a delusion that close relatives are impostors, is based on the impairment of affective tagging. The patient sees the image of the loved one, but it does not trigger the affective feeling characteristically triggered by the image. As a result, the patient does not emotionally recognize the person as his or her loved one, but as an impostor.

[258] Likewise, the experience of free will, by some considered as the illusory experience of being "in charge" [311], can be also approached from the perspective of tagging streams of information supporting motivation for action.

[259] The event-related potential Processing Negativity (PN) can be interpreted as the matching of top-down expectation and bottom-up input and the N200 as corresponding to a mismatch that resets the active category and initiates memory search for a better-matching recognition category, potentially representing a novel bottom-up stimulus [162].

[260] Reported by Koch and Crick as *"zombie behaviors"* [312].

[261] And the authors [161] warn, *"...top-down attention and consciousness, so often conflated, are not the same. One of the consequences of this distinction is that many of the neuronal correlates of consciousness that have been reported are probably confounded by the neuronal correlates of attention".*

[262] The mirror test of self-awareness utilized in non-human animals.

[263] In this context, "representation" is a pattern of information or encoding that denotes something; a pattern of neuronal activity, changing synaptic weight in network learning, and other neural states.

[264] Processes of consciousness are secondary processes to primary preattentive processes, an arrangement that implicitly supports a hierarchical order [313].

[265] Intention is a better term, but here I use it interchangeably with "command" for better clarity.

[266] More precisely, a target movement that is converted into control signals by an inverse model [61].

[267] *"The existence of independent phenomenal space is nowadays well confirmed by neurophysiological and cognitive investigations ...this phenomenal space may be identical with some aspect of brain space but not with any aspect of external physical space ...the pure phenomenological space – just an empty 3D matrix – resides at the level of brain organization that is ontologically subphenomenal (we never experience subjectively the countless coordinate system as such directly; we should know about it only through the relations among phenomenal objects)."* Fingelkurts and colleagues [54]

developed an Operational Architectonics (OA) framework to address this brain-mind interaction: *"OA theory explores the temporal structure of information flow and inter-area interactions within a network of dynamical, transient, and functional neuronal assemblies (whose activity is 'hidden' in the complex non-stationary structure of the EEG (and/or MEG signal) by examining topographic sharp transition processes (on the millisecond scale) in the EEG. Detailed analysis of the complex structure and hierarchical architecture of EEG reveals the existence of particular operational space-time (OST) which literally resides within the brain internal physical space-time (IPST) and is functionally isomorphic to the phenomenal space-time (PST)."* They found *"sequences of segments between different local EEG signals are indeed synchronized to a certain extent and form short-term metastable topological combinations (Operational Modules - OM), with different size (number of cortical locations involved) and life-span (temporal duration). We argue that at the phenomenological level, the lasting OM would be experienced as a 'phenomenal present' of consciousness. This hypothesis remains to be proven experimentally, however some empirical evidence already exists. For example, the mean duration of OMs usually varies from 80-100ms (for large OMs that cover a large part or indeed the entire cortex) to 30ms (for small OMs). These accounts, including variations in duration, are consistent with know estimates of cognitive processes, and of highly dynamic "moments of experience" or "thoughts", which may vary between ~100 milliseconds and several seconds depending on circumstances. At the same time, it was shown that some OMs are surprisingly stable and persistent across all studied experimental conditions in all subjects. We have demonstrated (data in preparation) that these highly stable OMs constitute the set of cortical areas that has been named the 'Default Mode Network' (DMN). Nowadays researchers tend to associate this DMN either with the stimulus-independent thought, mind wandering and the internal 'narrative' or with the 'autobiographical' self, 'chronic' self-evaluation, being a 'self', or having self-consciousness ...a subject that experiences multimodal perceptual phenomenal self-consciousness always feels directly present in the center of an externalized multimodal perceptual reality. This well-known (from the first person perspective) fact specifies that the sense of 'being a self' (probably in an implicit form) is always active even during the realization of any cognitive or other tasks, independently of their complexity and also in the dream state."*

[268] *"We suggest that our results point to a functional independence between the sensory modulation of oscillatory signals that are employed by control processes (beta oscillations) and conscious processing in the prefrontal cortex under baseline, control-free conditions. Furthermore, it is likely that beta oscillations could reflect an intrinsic mechanism of elementary control due to the pattern of modulation observed as a result of sensory input. Apart from higher-order processes, control functions can apparently engulf more basic functions that satisfy the criterion of disturbance compensation. Our results suggest that visual sensory input represents a disturbance to the cortical network interactions responsible for generating the intrinsic prefrontal beta rhythm. This sensory disturbance results in the initial desynchronization of beta oscillations as reflected in the beta power reduction. During that period the network interactions responsible for beta become destabilized and result in a reduction/desynchronization of beta power but soon control over this disturbance is achieved by the underlying network as reflected in the*

rebound of beta activity ~400 ms following a change in visual input. The similarity of this effect for PA and BFS, perceptual dominance and suppression, points to an independence of this elementary mechanism from the coexisting neuronal networks underlying conscious perception" [314].

[269] A number of brain regions display gamma oscillations, including isolated pieces of cerebral cortex. This specific brain rhythm emerges from the coordinated interactions between excitatory and inhibitory neurons. The pyramidal cells of the cortex use glutamate as their neurotransmitter, and only have excitatory activity. Approximately 80% of the afferent synapses that connect to pyramidal cells have excitatory activity. These synapses originate from other local and non-local pyramidal cells [174]. The remaining 20% of the synapses with inhibitory activity originate from inhibitory interneurons. In the absence of an inhibitory balance, the cortex would undergo runaway excitation and epileptic discharges. However, rising pyramidal excitation recruits inhibitory interneurons, creating a specific and tightly self-controlled cycle of excitation-inhibition, which gives rise to gamma oscillations [82].

[270] These patients passively listened to 110 repetitions of 2.5-second sentences. The authors found that the brain uses directional multiplexing. Multiplexing in telecommunications refers to multiple frequencies to transmit signals on a common physical support. In auditory processing [126] the term has been used to describe parallel information conveyed at different timescales.

[271] Delta-beta frequencies dominated the T-D direction, while gamma frequencies dominated the B-T direction. These results are consistent with a predictive coding framework; with B-U accumulation of prediction errors and discontinuous T-D prediction flow from nonlinear mapping of higher-level representations. In a more detailed description: very low frequencies at the delta range (1-3Hz) were implicated in the T-D direction. Delta phase modulated gamma activity at the primary auditory cortex, mostly at the left auditory cortex. The power of gamma activity was modulated by the phase of delta-beta activity from the T-D perspective, but not the other way around. Information did not transfer continuously, but by alternating T-D and B-U processing in periods of ~300-500ms. The slow alternation at 1- to 3Hz rate probably reflects time constants specific to the syllable-based mechanisms of speech. The authors [126] pointed out that in vitro models of beta-activity generation indicate that it can be the outcome of two gamma rhythms (of higher frequency) produced at superficial and deep cortical layers. The alternation between beta and two gamma rhythms conceivably serves as a switch device from upward information flow to its confining and redirection downward. Alternatively, alpha rhythm occasionally splits in high theta and low beta components or could generate beta oscillations by the interaction of feedforward signals.

[272] Hyafil's model is based on the intracranial recording of neuronal selectivity in human subjects. Individual excitatory neurons in the theta module received channel-averaged input, while, in the gamma module, neurons received frequency selective input [177]. The theta excitatory population "*conveys a widely tuned temporal signal capturing stimulus fluctuations*", whereas the gamma-excitatory population represents the "*rich spectral structure of the stimulus*". Note that these results parallels the relationship between fast gamma synchronization (coupling) by specificity-projecting tufted cells and slow gamma synchronization by dispersedly-projecting mitral cells in olfaction [149].

[273] Speech fluctuations phase-locked theta rhythm. Theta rhythm entrained gamma. Theta activity provided a referential syllable-based internal time necessary for reading the activity of gamma neurons: *"the structure enables syllable boundary detection (through the theta module) to constrain the decoding of faster phonemic information"*.
[274] Sensory de-multiplexing: the parallel processing of different frequency streams in sensory signals [177].
[275] A pure bottom-up approach limits Hyafil's conclusions. Top-down predictions play *"a significant role in guiding speech perception"* [177], as expected in predictive coding [178 - 180].
[276] In brain states characterized by oscillation low-excitability suppresses the processing of events. For this reason, situations demanding continued vigilance for possible events that can occur at any moment in time seem to be associated with suppression of underlying slow oscillations. In states of higher oscillatory excitability the probability of detecting an event is more contingent on the receiving oscillatory phase.
[277] However, there may be informational limits of gamma oscillations. According to Merker, the amplitude and spacing of the waves of excitation-inhibition that characterize gamma oscillations match membrane excitatory drive on a cycle by cycle basis, where amplitude and frequency change from one cycle to the next [174]. The inhibitory response to higher excitatory amplitudes brings about a longer interval between cycles, *"establishing a linkage between changes in amplitude, frequency and phase across individual gamma cycles"*. At a cycle length difference above 28ms, spanning "*the full 29ms phase range (2Pi radians) of the oscillatory means frequency of 35Hz*" [174]:

> "*The three principal dimensions or parameters characterizing an oscillatory phenomenon are frequency, amplitude, and phase. As just noted, the cellular events ensuring excitation-inhibition balance in the cortex are reflected in cycle by cycle changes in all three of these parameters of the local field potential gamma rhythm. With gamma amplitude, instantaneous frequency, and phase all tied up in the basics of balancing excitation with inhibition from moment to moment, no substantive degrees of freedom remain by which additional information dimensions - such as cognitive ones - might be reflected in or carried by cortical gamma*".

In other words, no information may bear an intrinsic connection to gamma oscillations. This kind of analysis can be helpful to delineate the oscillatory complexity capable of representations supporting qualia. Gamma oscillations constitute a marker of cortical tissue activation [174]. Thus, gamma cortical signals strongly correlate with bold fMRI signals, as well as with cognitive and task performances. This analysis did not argue against the transport of information through separate oscillators and through cross-frequency coupling of rhythms. As previously discussed, slower rhythms support and modulate gamma oscillations [82, 173, 181, 182]. Gamma-oscillations are short-lived. Transient cell assemblies organized into gamma-wave cycles [82] could allow for the realization of information across cellular assemblies. Computations carried by individual neurons were not part of the analysis, either. It is also thought that in the laminar cortical model of category learning, a *"good enough top down attentive match with a bottom up feature pattern yields faster gamma oscillations, whereas a big enough mismatch yields slower beta oscillations"* [162]. The gamma-beta dichotomy of match-mismatch dynamics indicates that gamma oscillations participate of information encodement by embodying a resonant matching state.

[278] For example, the activation of visual area 4 (V4) with color identification [186] and relating V5 for motion in specific paradigms [162, 163].

[279] For example, higher order representations in the posterior parietal regions and dorsolateral frontal cortex and first-order representations in the sensory systems [187].

[280] Aru and colleagues reviewed the methodology used in studies that addressed the neural correlates of consciousness [188]. Most of these studies applied contrastive analysis of brain functions. Contrastive analysis protocols maintain a similar target stimulus throughout the experiment, while the conscious perception of the target may or may not occur, *i.e.* it becomes the dependent variable. Thus, researchers probe into the neural correlates of consciousness by contrasting measurements of neurobiological activity between conscious and non-conscious trials. However, confounding factors may affect the results. Such factors include prerequisite local processes that facilitate neural correlates of consciousness and interfere with measurements. As I previously discussed, phase and oscillatory power may reflect local excitability, and affect the likelihood of conscious perception. Conversely, any brain activity that occurs shortly after neural correlates of consciousness can also affect the results of contrastive studies. For example, consequences of conscious perception, such as top-down control, report, or performance of an associated task, can all contaminate events that occur 300ms after stimulation, which researchers identify as manifestations of neural correlates of consciousness. Previous work has also indicated that the expectation posed by some experimental paradigms and after-image memory of the stimulus could have a fringe participation in neural correlates of consciousness, rather than simply representing a prerequisite or after-effect of conscious activity [194].

[281] The authors showed each subject images of four categories for 250ms. Between images, a gray screen appeared for 750ms. Subjects had to press a button whenever exposed to two identical consecutive pictures and should not do anything otherwise.

[282] BHF power of the local field potential reflects average neuronal firing. The authors used BHF as a measure of gamma oscillations (50-150Hz). Local field potentials (or wave packets) are the unified mean-field potential of membrane assemblies generated by the synchronized activity of thousands of neurons in the extracellular space of the cortical sheet [54]. They are understood to generate the EEG.

[283] Evoked response potentials (ERPs) yielded results that resembled those observed with brain broadband frequency. The authors found two ERP peaks (100-150ms and 250-350ms) instead of one in the visual cortex. Most electrodes showed a latency of less than 200ms, but there were enough with longer latency to produce those results. The overall peak latency of the evoked response potentials was ~300ms post-stimulus onset. They concluded that ERP activity in non-visual regions gradually increased, with a peak at ~300ms. They concluded that ERP activity is likely the dominant factor affecting the scalp EEG response peak at ~300ms [201].

[284] The authors speculated that the ultra-fast "glow" informed non-relevant cortical regions of a high activation in content-selective areas.

[285] Cross-modal recruitment.

[286] *"Individual cortical neurons and their probabilistic firing can simultaneously participate of multiple functional neural ensembles. That means that the spikes produced by a single cortical neuron can be utilized by distinct neural ensembles to encode multiple functional and behavioral parameters. Thus, even though in a particular moment in time a single cortical neuron may exhibit sharper tuning to a particular motor or sensory parameter, its spiking may concurrently participate in the encoding of a different parameter, performed by another subset of neurons. The neuronal multitasking principle predicts that the entire cortex is capable of cross-modal sensory responses, and that individual neurons are capable of encoding multiple motor and higher cognitive functions."* [164, 207].

[287] And it blurs the distinction between 1st order and higher-order representationism.

[288] Models for the integration of brain activity layers bring about pertinent questions [182]. Slow oscillations associated with cardiorespiratory activity seem to modulate the excitability of the resting state network. Infra-slow oscillations (0.01-0.1Hz) can modulate faster oscillations [208]. Infra-slow oscillations strongly correlate with faster oscillations in the range of 1 to 40Hz [209]. Slow oscillations provide the basis for subsequent faster oscillations. A first layer of human brain activity emerges from the interactions between these slow hemodynamic and respiratory oscillations and the default mode network (DMN) oscillators [182, 210]. The DMN seems to play a fundamental role in consciousness and neural connectivity. It utilizes most of the energy consumed by the brain [211]. It integrates internal and external information by creating an oscillatory "framework" for processed information to reach conscious awareness [182]. A second layer of human brain functional organization is thought to consist of limbic oscillations and their associated cardiorespiratory oscillatory activity. Cortico-thalamic oscillations may constitute the third, and higher, oscillatory layer [182]. Jerath and Crawford suggested that delta oscillations integrate homeostatic processes by coupling with slow oscillations [182]. Beta and theta oscillations were considered to characterize the second layer of activity, and alpha and gamma oscillatory activity associated with higher cognitive processing. This hierarchical cross-coupling was seen as generating an internal 3D default space composed of all interconnected cells in the body, thus *"recreating the external world within the mind"* [210].

CHAPTER 17

The Global Neuronal Workspace Theory and the Adaptive Resonance Theory

Abstract: The Global Neuronal Workspace (GNW) theory addresses how the brain enables the coherent self-sustaining states characteristic of conscious experiences. Corticocortical long-range fibers make information available for multiple brain networks through horizontal projections. These neurons are considered the distributed source of a Global Neuronal Workspace. The GNW differs from the IIT because it does not situate integrated information as the cause of consciousness. Instead, it points towards information-sharing mechanisms as the origin of consciousness. A recent update to the theory stresses the functional unity of cortex and thalamus. Information integrated by the cortico-thalamic (C-T) circuitry generates internal representations of the external space through the widespread synchronization of gamma oscillations. The theory also proposes mechanisms to generate abstract states of awareness not traditionally associated with qualia. The Adaptive Resonance Theory (ART) describes dynamic states that the brain amplifies and synchronizes during the matching processes between bottom-up and top-down streams of information. The ART predicts that qualia should emerge from the emergent relational properties of these interactions. High-level resonances that support qualia features are described as shrouds and presented as stable invariant dynamic states. As a net effect, free-energy, or uncertainty, decreases across hierarchical systems. The ART fits the concept of conscious experiences as the abstract realization of information by strange loops, connecting top-down and bottom-up information flows. ART shrouds are consistent with the manipulation of information as focal aspects of subjective experiences, such as the odor-objects. Matches between prediction and predictive-error create perceptual representations of the world that can support integrated, goal-directed adaptive behaviors of the organism.

Keywords: Adaptive Resonance Theory, Corticocortical Fibers, Corticothalamic Fibers, Cortico-Thalamic Core, Cognitive Heuristics, Cross-Frequency Coupling, Feelings of Knowing, Focal Aspects of Subjective Experiences, Free-Energy Reduction, Global Neuronal Workspace Theory, Higher Order Representationism, Predictive-Error, Primitives, Reentrant Signaling, Shroud-Resonances, Stability-Plasticity Problem, Thalamus, Top-Down and Bottom-Up Interaction, Winner-Take-All Coalitions.

Paulo J. Negro

THE GLOBAL NEURONAL WORKSPACE THEORY

Experimental evidence suggests that human conscious experiences emerge from globally distributed brain activity, and to intense and correlated activity among distant cortical regions [46, 47, 61, 183]. The Global Neuronal Workspace (GNW) theory [183, 190, 191, 205, 212-214] incorporates this body of evidence in one of the most accepted theories today. Long-range excitatory fibers originating at the prefrontal, cingulate, and parietal cortices, as well as thalamocortical loops sustain a neuronal workspace that gave rise to human consciousness. These long-range pathways are thought to enable coherent self-sustaining states consistent with conscious experiences [215 - 218].

The global workspace network contains a large number of interconnections among the dorsolateral cortex, the inferior parietal cortex, the mid-temporal cortex, and the precuneus [215 - 218]. Specialized modules, such as processors for language, sensations, and reasoning, among others, compete for access to the global workspace. Coalitions of these local processes gain access to the global workspace and broadcast emerging messages back and forth from their cohort of parallel, specialized local networks.

Consciousness is seen as emerging out of the collective dynamic of these agents playing into this widespread network [61]. The brain interprets locally received signals, uses them to update local processes, and refreshes these processes with reentrant signals that go back and forth from their source [218]. Local information processed by parallel specialists is not conscious, whereas broadcasted information to the whole cohort is consciously experienced [215] (Fig. **12**).

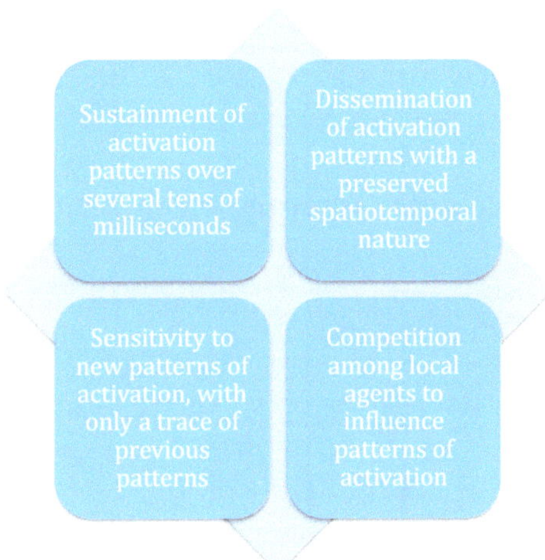

Fig. (12). Principles verified by modeling of a simple, but heuristic, version of the GNW [213].

The GNW theory differs from the IIT in pointing global information sharing as the origin of human consciousness [189,215-218]. According to the IIT, consciousness equals integrated information. On the other hand, GNW theory proponents defend that consciousness emerges from a neural architecture characterized by flexible information sharing throughout the cortex, widespread integration and broadcasting[289]. Here, integration of information does not play the role of consciousness primary cause[290].

There is no single anatomical hub in the GNW. Corticocortical long-range fibers make information available to multiple brain networks through horizontal projections from the pyramidal cells of layer 2 and 3 [218]. These neurons are considered the distributed source of a Global Neuronal Workspace. Consciousness emerges from multiple functional hubs that bind and propagate neural signals across task-related networks, in a winner-take-all dynamic [189]. This dynamic supports the global integration of functional networks and large repertoires of functional activity states [215 - 218].

In fact, network simulations indicate massive unconscious processing interspaced with serial conscious stepwise decisions that route information among specific processors [46, 47]. Electroencephalogram (EEG) studies show a late (~300ms) event-related potential (P300) marking the experience of perceptual consciousness [219]. High-density EEG data indicate that strong local interactions and robust interconnectivity underlie consciousness [219, 220]. Thus, conscious experiences correspond to information within a GNW, represented by stable brain-scale assemblies and organized by a system of rhythms.

A BRIEF REVIEW OF THE THALAMIC PHYSIOLOGY

The thalamus is an egg-shaped structure located at the diencephalon[291]. It relays ascending signals to the cortices, gatekeeping information according to behavioral states. Many thalamic nuclei[292] exist: some receive information regarding specific sensory modalities and channel this information to cortical areas; others project widely to the cortex and subcortical regions [221]. The thalamus receives prominent projections from the cortices. The structure mediates cortical arousal, as well as sensory information integration[293].

Information integrated by the thalamus is supposed to generate an internal representation of the external space through the widespread synchronization of gamma oscillations [205,212]. Proponents of the GNW theory recognize the crucial role of the thalamus as a fundamental part of what they designated a Cortico-Thalamic (C-T) core [189,218] of the dynamic Global Workspace (dGW) extension of the theory[294].

The C-T core "parallel-interactive" organization follows a small-world architecture[295]. Highly integrating and differentiating capabilities characterize small-world functional architectures[296]. C-T circuits embody "any-to-any" connections, bidirectional signaling[297], and the participation of specific cells across multiple activity arrays[298]. This kind of organization allows the overlay of multiple "spatiotopic"[299] maps.

Through the cooperation and competition of activity streams across these maps, conscious percepts are thought to emerge. Their emergence is the result of "winner-take-all" reentrant[300] coalitions [189]. In EEG recordings, these coalitions appear in association with conscious stimuli as content-sensitive stable microstates of 100 to 200ms (theta oscillations) [193].

However, a word of caution is warranted.

Experiences such as self-referential thinking, remembering and imagining the future, social cognition, mind-wandering, theory of mind, spontaneous cognition, mental time travel, and the internal train of thought *"have one thing in common"* [176]: they are not specifically associated with thalamic activation. These experiences are characteristic of the Default Mode Network (DMN).

PARALLELS BETWEEN SILICON AND BRAIN COMPUTATION

Theories of consciousness share the idea of consciousness realization in neural states [163], where each conscious state corresponds to a specific neural state. However, theories of consciousness should not eschew computational hierarchies. Information must be realized in abstract levels.

The GNW theory implicitly encompasses the idea of realization of information. According to it, conscious perception emerges from *"frame binding"*, described as arrays that do not give rise to conscious experiences while providing a scaffold for conscious percepts. However, the theory does not emphasize enough the hierarchical organization of the brain[301], a key architectural principle. At the center of the brain's perceptual inference resides the recurrent dynamics of predictive-error suppression, across multiple hierarchical levels [222] as described by the Adaptive Resonance Theory.

Silicon-based architectures depend on *"the abstraction of lower-level realization of computation in hardware through an elaborate succession of more abstract descriptions of the same in levels of software"* [40]. Conceptually, brain computation should not be different: increased abstraction levels allow for greater spatiotemporal scales and minimization of underlying computational complexities.

The brain is organized in different spatial scales: molecules, synapses, neurons, networks, maps, systems, and the brain itself. Codes become more and more abstract, eventually reaching a "qualia size" that plays a role in behavioral goals. Despite already discussed localizationism misgivings, information has an embodied origin, *i.e.*, a known origin: the "labeled-line" concept used in silicon computation and built into the spatiotopic activity maps. The overlay of spatiotopic maps is not conceptually different from the overlap of omnidirectional hippocampal cell assemblies participating in higher order representations. New relational properties can emerge in both systems from nothing of their own kind.

From a computational perspective, the C-T system can be seen as retrieving and recoding cortical information. This process finds a parallel in silicon computing, when a single processor executes a sequential list of instructions. Metaphorically, the C-T system "googles" table-lookup organized cortical memories, and each state of the cortex represents a previously indexed Web page that the brain quickly accesses [40]. Importantly, the cortex-thalamus unity, unlike a silicon-based driver program shuffling input and output in standard computation, utilizes a compressed form of expectation-based coding that signals the differences between what was expected to occur and what actually occurred.

Predictive coding refers to the dynamics of perceptual inference driven by top-down priors from levels above to levels below,

> *"...recapitulated at all level, enabling self-organization through recurrent interactions to minimize free-energy and suppress predictive errors throughout the hierarchy. In this way, higher levels provide guidance to lower levels and ensure an internal consistency of inferred causes of sensory input at multiple levels of description" [72].*

Simply put, this is the description of a subject realizing information in a progression of representations with different levels of complexity. This view supports that theories of consciousness embrace the idea of consciousness as generative action. Theories of consciousness must necessarily acknowledge the existence of a subject from the beginning[302].

GLOBAL AND LOCAL CONNECTIVITY AS A FORM OF INFORMATION

After the stimulation of sensory regions, changes in early local excitability do not differentiate between conscious and unconscious perception. However, pre-stimulus excitability greatly affects whether the subject will consciously experience a near-threshold stimulus[303] [202, 223]. Alpha oscillations function in the top-down control of local networks, likely serving as a gating mechanism that inhibits task-irrelevant regions and modulates the "expectation" of near-threshold stimuli[304] [94].

Alpha activity can pre-activate functional networks, differentially integrating different sensory areas prior to stimulus presentation [224]. However, alpha power does not equal functional connectivity. Pre-established connections may trigger associations among different networks[305] even though top-down influence may enhance global small-worldness and local clustering and efficiency [94,225]. It is not only local conditions that predict stimuli perception, but also ongoing pre-connectedness among local sensory areas with the essential nodes of the global workspace.

SUBJECT-CENTRIC SYNCHRONIZATION

Reentrant thalamocortical and corticothalamic oscillations can process information in a format that allows for conscious access [205]. Cross-frequency phase coupling integrates the full range of cortico-thalamic rhythms[306] [189, 224-226]. In a review of the methodology applied in functional brain studies, Uttal stressed that meta-studies typically identify a broader distribution of activated regions than individual studies[307] [227]. These results indicate that conscious adaptive experiences involve the collaboration of many regions in many organizational levels [228]. Methodologies that minimize fMRI false-positives may underestimate widespread task-locked activation. This should not come as a surprise as the brain is in different degrees continuously active, in all its parts [73, 229].

ABSTRACT STATES OF AWARENESS DECOUPLED FROM QUALIA

Information flowing from sensory cortices to frontoparietal regions has a clear figure-ground contrast. The perception of a cup of coffee clearly detaches from the background in terms of vision, smell, and taste. Conversely, information flowing from the frontal regions has poor figure-ground contrast.

The GNW theory proposes mechanisms that generate abstract states of awareness, such as feelings of knowing (FOK), in addition to states of awareness traditionally associated with qualia, such as vision and color. Information is considered to move anterior-posteriorly in the more abstract awareness and posterior-anteriorly in the qualia laden awareness.

Coalitions that bind and broadcast FOK, semantic knowledge, and other abstract concepts fall into the anterior-posteriorly information flow. Those abstract states of awareness support parsing the Hard Problems in two aspects, one of them being a mechanism for realization of information.

As previously reviewed, Kahneman [118, 119, 206] System 1" allows for fast processing of information, largely without a sense of voluntary control, and a "System 2" allocates attention to effortful mental statistics. Kahneman is focused on how the human mind "thinks" about choices and calculations or, in other words, the heuristics of the abstract short-cut bias of human reasoning.

System 1 "thinking" is based on implicit causal narratives that support hypotheses about the world. These narratives are ecologically adaptive. They suggest the evidence of mechanisms for a Bayesian sampling of the environment. [120, 121]. These inferential narratives capture abstract relationships that go beyond a strict phenomenological description.

THE ADAPTIVE RESONANCE THEORY

According to the Adaptive Resonance Theory (ART), dynamic states of resonance [308] mediate the interaction between top-down expectations and bottom-up sensory information. Qualia emerge from these networks interactive properties. Resonances are thought to exhibit the parametric properties of individual conscious experiences when *"embodied within the spatiotemporal patterning of cell activities across networks of feature detectors*[309]*"* [162].

Pairs of complementary cortical processing streams[310] interact to generate emergent properties overcoming their complementary deficiencies[311]. A small number of equations describe these computational modules[312]. Resonant dynamic states support qualia through the generation of abstract emergent properties of a relational origin.

The theory provides a framework for consciousness as an abstraction: *"cycles of resonance and reset underlie much of the brain's perceptual and cognitive category learning dynamics"* [162]. These cycles can be seen as supporting the realization of information into abstraction through strange loops by providing the degrees of freedom required for emergent levels of awareness:

> *"...laminar circuits of the cerebral cortex seem to realize a revolutionary computational synthesis*[313]*of the properties of feedforward and feedback processing, digital and analog processing, and data-driven bottom up processing and hypothesis-driven top-down processing".*

The brain relies on multiple reprocessing to generate complete and stable representations of the world, to control adaptive behaviors and to ultimately *"mark these representations with conscious awareness"*. The net effect of this activity is a minimization of uncertainty across hierarchical processing that resembles the reduction of free-energy or surprise.

Reviewing the ART in detail would go beyond the scope of this work. However, I will stress a crucial issue for the theory: not all brain states are resonant and not all resonant states are conscious, but, according to the ART, all conscious states are resonant. In other words, cells that discriminate the characteristics of qualia-experiences necessarily sustain resonant states during those conscious experiences[314]. Moreover, these conscious states only occur at higher-order levels of brain processing.

High-level resonances that support qualia features are described as "shrouds"[315]. The synchronization of these shroud-resonances supports comprehensive conscious experiences. For example, "stream-shroud" and "spectral-pitch-and - timbre" may synchronize during conscious hearing and during the recognition of auditory streams [162].

Shrouds emerge as stable invariant states that compensate, say, for variable illumination of visual input, generate the illusion of auditory continuity, sustain parallel auditory streams, intra-word and inter-word invariance, and decrease uncertainty towards a final hierarchical resolution[316]. ART shrouds are consistent with the manipulation of information as focal aspects of subjective experiences, such as odor-objects reported in the olfaction chapter.

The ART implicitly relies on actions and explicitly on oscillatory state matching. It is a relational model that attempts to explain self-stabilizing representations of objects and events. Resonant dynamic states support qualia through emergent properties of relational origin. These states are the product of cycles of feedback and feedforward processing. Multiple reprocessing of information ultimately marks representations with conscious awareness. The net effect is a reduction of free-energy or uncertainty across hierarchical structures. The ART thus fits the concept of conscious experiences as abstract realization of information by strange loops, connecting top-down and bottom-up information flows.

CONCLUSIONS

The Global Neuronal Workspace and the Adaptive Resonance theories are not incompatible. They describe horizontal and vertical mechanisms supporting consciousness. Resonant states can underlie:

'expectation.information'expectation.information

oscillations capturing high-level synthetic descriptions of perceptual objects. An active expectation is considered a "prediction" and information is seen as "predictive-error". Processing of information as conscious experiences encompasses multiple operational levels because *expectation.information* of lower levels is taken as a *primitive* and manipulated as "information" at higher levels, as depicted at (Fig. **2**).

Realization of information as the hierarchical violation of a strange loop can be made possible when *expectation.information* ceases to be taken as a primitive:

'expectation.information' ⟷ *expectation information*
'prediction.predictive-error' ⟷ *prediction.predictive.error*
'feedback.feedforward' ⟷ *feedback.feedforward*
'top-down.bottom-up' ⟷ *top-down.bottom-up*

(at the same computational hierarchical level)

within the strange loop:

'prediction.predictive-error'prediction.predictive-error
'feedback.feedforward'feedback.feedforward
'top-down.bottom-up'top-down.bottom-up

It is intuitive to think that the matching of prediction and predictive-error identifies those perceptual representations of the world capable of supporting the organism integrated, goal directed adaptive behaviors. However, the enactment of global adaptive behaviors in an unpredictable environment does not allow for precise calculations of predictive-error of those phenomenal representations. At this level of representation, Bayesian computational limitations favor the transition into phenomenal and cognitive heuristics.

NOTES

[289] The IIT considers global broadcasting of conscious information necessary, but primarily as a tool to create an integrated representation of its various features. The GNW supports a more directional broadcast of bound conscious content [189].

[290] This view is consistent with the "Neural Darwinism" theory of neuronal group selection. Neuronal Darwinism is a theory of neuronal group selection serving the diversity and complexity of adaptive behaviors. Neural Darwinism neural selection follows three principles [218]:

1. Selection of a primary repertoire during brain development: neurons firing together wire together
2. Selection of a secondary repertoire by synaptic strength during and after development: selection based on individual experience;
3. Dynamic adaptation by reentry of long-range, reciprocal and massively parallel connections: support of dynamic spatiotemporal relationships for the integration of adaptive conscious behaviors.

[291] Diencephalon: it contains structures lateral to the third ventricle, including the thalamus hypothalamus, epithalamus and subthalamus.

[292] Most thalamic neurons use the excitatory neurotransmitter glutamate. The reticular

nucleus, a sheet-like structure covers the thalamus. The reticular nucleus does not project to the cortex. Most of its neurons deploy the inhibitory neurotransmitter ϒ-aminobutyric acid (GABA). Fibers that leave the thalamus project collaterals to reticular neurons while crossing the reticular nucleus. Reticular neurons modulate the thalamocortical flow of information like a filter [221].

[293] As reviewed by Jerath, Crawford & Barnes [210], damage to the thalamus can lead to grave complications, including memory deficit, sensory-motor dysfunction, speech impairment, and altered states of consciousness, such as coma or stupor. Conversely, central thalamic stimulation can restore consciousness in patients in coma or in a vegetative states. The thalamus does not play a role solely in exteroception. It also takes part in attention, interoceptive and otherwise self-referential activity. Thalamic injuries can cause severe impairment of functional integration, regulation and arousal. They can also cause epileptic seizures. Thalamic infarcts can produce secondary generalized decrements in cortical metabolism.

[294] Multiple codes are available to the C-T core, including signal propagation through glial cells (glial cells are non-neuronal cells supporting the neurons), electromagnetic field induction, electric synapses, and membrane ion currents [189]. The core was characterized as the "*oscillatory machine*" sustaining the spontaneous rhythms associated to behavioral states. Its functional unit is neither a single neuron nor a unidirectional sensory pathway, but units of "*adaptive resonance*" formed by the functional connections between cortical and thalamic neurons.

[295] Corticothalamic connections are thought to support spatial superposed line coding of information, which preserves relative locations of neurons in maps. "*At higher levels, C-T arrays project to self-centered and object-centered spaces, which must be coordinated with each other to support a coherent domain of action and perception... Cortical arrays therefore resemble the head-up cockpit display of a fighter jet, which may show a geometric frame for targets, and more for attackers and neutrals. When these floating spatial frames intersect, they may alert executive networks to make decisions*" [189].

[296] This architecture is different from the cerebellum, which is also organized in a parallel structure, but without the interactions emerging from bidirectionality and the any-to-any connections of the C-T core.

[297] Bidirectional connections: corticothalamic, thalamocortical, and corticocortical pathways that provide the structure for a reciprocal signaling integrating posterior modality-specific cortical areas and anterior portions such as the prefrontal and cingulate regions [218].

[298] In the dGW theory, conscious perception is thought to result from the interactions among spatiotopic maps. These interactions involve *"multiple overlays, like map transparencies."* Baars and colleagues reported the existence of higher-level C-T arrays that project to self-centered and object-centered spaces, thus coordinating action and perception [102]. Conscious perceptions that emerge from winner-take-all coalitions would then propagate signals across these multiple arrays: *"...high-level wave-like phenomena in the C-T system, including standing and traveling waves, spiral vortices, centrifugal propagation, phase coupling and decoupling, microstates cross-frequency coupling and hemisphere-wide phase transition at theta-alpha rates"* [189].

[299] Spatiotopic: referring to three-dimensional space of the external world.

[300] The emphasis on the C-T core marks the transition of the GNW theory into a dynamic Global Workspace theory. According to the dGW theory, conscious experiences emerge from *"reentrant signaling among topographical regular spatial maps"* [189]. Reentrant signaling does not constitute a simple form of feedback; it provides a mechanism for the selection of *"trajectories"* in the multi-layered two-dimensional neuronal arrays of the C-T core.

[301] The theory does not dismiss hierarchical organization, but rather emphasizes space-time assemblies of information. According to Baars and colleagues, task-specific activity results from *"cross-frequency coupling at multiple spatial scales, linking and unlinking multiple sites in the C-T core and its satellites"* [189].

[302] Most authors do not incorporate these ideas in the interpretation of work on neural correlates of consciousness (NCCs) and the global neural workspace. Instead, their interpretations of empirical evidence have an intelligent design bias. For research protocols specifically addressing NCCs, actual results are unlikely to be directly affected. For example, a meta-analysis of neural correlates of visual consciousness found clusters of activation in the inferior and middle occipital gyrus, fusiform gyrus, inferior temporal gyrus, caudate nucleus, insula, inferior, middle and superior frontal gyri, precuneus, inferior and superior parietal lobules [315]. These results are consistent with the GNW model. However, this paradigm biases theories of consciousness. A model that stresses the realization of information across hierarchies and perceptual domains will likely predict that conscious experiences should result from a more global activity than observed in protocols that target the identification of specific NCCs.

[303] Pre-stimulus, near-threshold experiments show that local, mostly sensory, areas with higher excitability prior to the detection of the near threshold stimuli are more likely to trigger a conscious response. As previously discussed, increased local excitability due to reduced levels of alpha-power (8-12Hz) in relevant sensory areas also predicts whether the brain will consciously perceive a given stimulus.

[304] Most studies show that conscious experience of a near-threshold stimulus requires the engagement of frontoparietal regions. However, pre-stimulus brain activity can determine whether the subject will consciously perceive stimuli [94,220,224]. Pre-stimulus alpha power and alpha phase modulate local excitability. When a near-threshold stimulus arrives, it "*will encounter 'pre-established' nonrandom routes, reducing the degree-of-freedom of information flow and determining stimulus future behavioral fate.*" The effects of attention alone do not explain this sequence of events. Enhanced connectivity to higher-order brain regions in the global neuronal network seems to constitute a "window" to conscious access that precedes stimulus arrival and fits with a predictive coding model of perception.

[305] In the posterior cingulate region, considered a core area of the default mode network, connectedness with the somatosensory cortex is associated with stimuli perception [94]. Auditory awareness is also preceded by regional excitability changes and enhanced integration of relevant sensory areas [223].

[306] Brain-body interactions seem to form a complex system that couples and decouples with a specific harmonic structure [224]. The dynamic of neural absorption and resonance has complex characteristics [225]. The coalescence of brain electrical activity in unified rhythms apparently results from cortically generated slow oscillation binding of other

rhythms through corticocortical and corticothalamic drives [226]. Some of these oscillations arise from intrinsic neuronal properties, but the synchronization of low frequencies and fast rhythms in complex wave sequences likely depends on the operations of the C-T system. The dynamic Global Workspace theory proposes that conscious contents emerge from the functional hub created by the C-T system, with its dynamic capability to bind and propagate signals across a small focal channel broadcasting every 100 to 200ms [189].

[307] As an example of wider activity, Gonzalez-Castillo and colleagues detected activity in most areas of the brain using bold fMRI during a visual stimulation plus letter/number discrimination task [229].

[308] Resonance: *"a system which is characterized by its own oscillation frequency and can enter into vibration if stimulated (subjected to sound waves, electromagnetic waves, or mechanical vibrations according to the nature of the system) by frequencies close to those peculiar to the system itself"* [54]. In these states *"...neuronal firings across a brain network are amplified and synchronized when they interact via* reciprocal excitatory feedback signals during a matching process that occurs between bottom up and top down pathways." Resonances are considered "adaptive" because they trigger network learning within adaptive weights and/or long-term memory. Reentry is considered a central mechanism for integration of brain function [197].

[309] *"...top down expectations focus attention on salient combination of cues, called critical feature patterns ...In particular, when a good enough match occurs between a bottom-up input pattern and a top-down expectation, a synchronous resonant state emerges that embodies an attentional focus. Such a resonance is capable of driving fast learning that incorporates the attended critical feature pattern into both the bottom-up adaptive filters that activate recognition categories, and the top-down expectations that are read-out by them – hence the name adaptive resonance – while suppressing unattended outliers that could have caused catastrophic forgetting."*

[310] The theory is based on complementary computing, such as the complementary processes of visual "boundary-surface", "boundary-motion", the "what" of perception/recognition and "where" of space/action, object tracking and flow navigation, and motor target position and volitional speed.

[311] The ART proposes a solution for the "stability-plasticity problem," the human brain's ability to learn quickly and stably without *"catastrophically forgetting past knowledge"*.

[312] Computational modules such as shunting on-center off-surround networks, gated dipole opponent processing networks, associative learning networks, and spectral adaptively timed learning networks: *"...an integrated self is possible because it builds on a shared set of equation modules within modal architectures that can interact seamlessly together."*

[313] Note the parallel with the "synthesis" of smells described in the chapter of olfactory awareness.

[314] *"It cannot be overemphasized that these resonances are not just correlates of consciousness. Rather, they embody the subjective properties of individual conscious experiences."*

[315] Shrouds such as "surface-shroud" visual resonances, entorhinal-hippocampal navigation resonances, "stream-shroud" auditory resonances, "spectral-pitch-and-timbre"

auditory source recognition resonances, “item-list” speech and language resonances, and “cognitive-emotional” resonances.

[316] The ART shroud-resonance provides a model of reconciliation for the local and global interaction paradox and a bridge between localizationism and distributivism.

CHAPTER 18

Networks of External and Internal Awareness and the Neurology of Consciousness

Abstract: The Default Mode Network is activated when there is no engagement with external experiences, but in experiences of self-awareness and endogenous consciousness. The DMN anticorrelates with Task-Positive networks, such as networks for executive control, stimulus salience, external attentional control, and external perceptual processing. The connectivity of a core self-awareness system overlaps with the DMN, in particular with the posterior cingulate cortex. Neural networks that produce social perception also overlap with the DMN. However, neither Task-Positive networks nor DMN computations directly generate a subject. The thalamus is not part of the DMN. Theories emphasizing the DMN and the GNW theory can be reconciled through phenomenological descriptions reported in classic Zen texts. Anesthesia studies point towards the disruption of signaling at upper cortical layers by general anesthetics with different pharmacological mechanisms. General anesthetics seem to act by disrupting top-down signaling and disabling mechanisms matching top-down and bottom-up informational streams at the apical dendrites of cortical pyramidal cells. According to proponents of the Orch OR, quantum phenomena in tubulin have a causal role in the regulation of synaptic firing. The theory suffers from the same biases of the IIT, as both propose a bottom-up causal origin of conscious experiences. Instead, research with general anesthetics suggests a crucial role for top-down projections and for the signaling that matches top-down and bottom-up streams. Children born without a cortex can display behaviors associated with consciousness. They can smile and laugh, show initiative and instrumental behaviors. Their parents can easily distinguish episodes of loss of consciousness due to absence seizures. In the damaged or cortically deactivated brain, secondary effects in the midbrain contribute to the loss of consciousness. The upper brain stem may support sustained rhythms encoding predictive coding expectations. Subcortical regions can provide an interface with the essential attributes of phenomenal consciousness. Phenomenal experiences may conceivably emerge with the onset of sustained oscillatory rhythms depicting sustained expectations in an integrative action-based model. Traces of phenomenal consciousness may thus emerge with the onset of rhythms that sustain an organism's internal expectations. These observations are consistent with the blueprint presented here.

Keywords: Anencephalic Children, Animal Consciousness, Cortical Deactivation, Coma, Cross-Frequency Coupling, Default Mode Network, Dendritic Signaling Disruption, Gamma Oscillations, General Anesthetics, Internal Expectations, Ketamine, Minimally Conscious State, Orch OR Theory, Phenomenal Consciousness, Propofol, Predictive Coding, Pyramidal Cells Apical

Paulo J. Negro

Dendrites, Quantum Phenomena in Tubulin, Sprague Effect, Superior Colliculus, Task-Positive Networks, Top-Down Projections, Zen Phenomenology.

THE DEFAULT MODE NETWORK

The brain uses most of its energy for ongoing spontaneous functions rather than the performance specific activities [229, 230]. These high energy expenditures are dedicated to the Default Mode Network (DMN) [211,229]. External task performance requires less metabolic activity than seemingly passive states associated with DMN activation.

The DMN is not composed of sensory or motor areas. It is isolated from a direct connection with the external environment, posing as the neural correlate of the streams of endogenous consciousness [173]. It is active with experiences of internal awareness and self-related thinking [211,231] but deactivated during goal-directed cognition.

The major DMN nodes are strongly interconnected, and include the medial prefrontal cortex, the posterior cingulate cortex (PCC), the insular cortex, the inferior parietal lobule, the lateral and inferior temporal cortex, the medial temporal lobes, and the precuneus [219,231-235].

The DMN participates in memory recollection and imagining the future through connections with the hippocampus and middle occipital gyrus. Dorsal PCC interactions with the prefrontal cortex may control attentional focus [231] across internal and external dimensions. In fact, one of the proposed roles for the PCC is the coordination of these mental representations.

The DMN has an inverse relationship with other large-scale networks. Systems activated by internally-directed (DMN) and externally-directed (attentional) cognitions are not only poorly integrated between themselves[317], but anticorrelate their activities in a variety of functional research paradigms and seem to switch their dominance every 20 seconds [231] (Table **5**).

A study of effective connectivity [233] in patients with loss of consciousness identified the PCC as a main hub of consciousness. The PCC was identified as a central recipient of excitatory afferents, and the origin of negative efferent connections. In minimally conscious and unconscious patients, afferent connections from anterior and parietal regions to the PCC also shut down the PCC inhibitory efferent feedback.

Table 5. Task-positive and Default Mode Networks [231].

NETWORK	REGIONS ACTIVATED	NETWORK SUBDIVISION
Default Mode (internal)	Midline anterior cingulate/mesiofrontal areas & posterior cingulate/precuneal cortices	
Task-positive (external)	Inferior parietal sulcus, inferior parietal lobule, precentral regions, dorsal lateral prefrontal cortex, V5/MT, insula, frontal operculum, supplementary motor area	Dorsal (top-down) attention Stimulus salience Executive control Frontoparietal perceptual

In that study, loss of consciousness paralleled a loss of connectivity between the PCC and other DMN regions, and stronger less discriminative oscillations within the PCC[318] [233]. In brief, a core self-system [176] composed of the PCC and the inferior parietal lobule may be gated to awareness by the medial prefrontal cortex (MPFC).

Human consciousness likely has a critical dependence on neural networks that promote social perception [236]. Construction of the content on other people's minds is associated with the activation of the superior temporal sulcus, the temporoparietal junction and the medial prefrontal cortex. The basic task of social perception is to reconstruct the focus of someone else's attention, requiring top down frontoparietal attentional control. From this perspective, human consciousness seems to emerge from neural networks that build up a theory of the mind with different tags for other people and the self.

However, neither Task-Positive networks nor the DMN carry out computations that directly create a subject. I have pointed out that the subject must be considered a noncomputational emergence. Nonetheless, these networks are the source of generative computations of external and internal experiences. For this reason, I see the absent subject as the active realization of information in both networks, depending on the nature of the experience, internal or external.

RECONCILING THE GLOBAL NEURONAL WORKSPACE AND THE DEFAULT MODE NETWORK

Although the frontoparietal system reported by the GNW partially overlaps with the DMN, corticothalamic signaling is not identified as a characteristic signature of the DMN. Nonetheless, the disruption of the connectivity of the posterior parietal and cingulate cortices, and of the nonspecific thalamus can both cause loss of consciousness [237]. Nonspecific thalamic nuclei are involved in the regulation of cortical arousal, the integration of information. Thus, these structures are thought to enable consciousness. However, the description of a biological mechanism may not capture phenomenal subtleties of conscious experiences.

12th century teachings of Zen Master Hongzhi Zhengjue expressed the meditative insight and the working of the awakened mind[319]. Traditional meditative practices involve concentration on specific objects or concepts in order to develop heightened states of concentrated awareness called samādhi [238]. According to the Zen scholar that translated Master Hongzhi's writings, *"... 'silent illumination', however, involves withdrawal from exclusive focus on a particular sensory or mental object to allow intent apprehension of all phenomena as a unified totality. This objectless meditation aims at a radical, refined nondualism that does not grasp at any of the highly subtle distinctions of our familiar mental workings.*"

Modern applications of meditation emphasize hypoarousal and relaxation. However, meditative practice is associated with increased brain alertness as evidenced by increased frontal theta power in highly focused cognitive states, increase gamma power associated with increased cortical processing, decrease in the DMN activity, and improved sensory and attentional processing. Meditative practices promote greater wakefulness and lower sleep propensity, particularly as practice progresses [239]: *"awakening is not a metaphor"*.

"*Opening the hand of thought*" is at the heart of Zen practice [240]. Master Hongzhi teachings emphasized a *"bright, boundless field"* of consciousness that carries no image or depends on anything[320]. I suggest that the regulation of cortical arousal goes beyond its biological mechanism or the idea of a quantitative level of consciousness. Cortical arousal as conveyed by thalamic interaction may underlie a qualitative, but subtle *"empty field"* of consciousness concealed under phenomenal external experiences and the internal narratives of the DMN.

Nevertheless, these ideas are controversial. It is argued that thalamic activity is not necessary *"for stimulus independent thoughts or specific endogenous streams of consciousness"* [173]. This view indicates that cortical dependence on the thalamus does not imply a thalamic role for conscious experiences[321]. This was made clear in the chapter addressing olfactory experiences as the olfactory system mostly bypasses thalamic relays. Nonetheless, contrastive studies may miss thalamic activity in perception or resting state paradigms. General anesthesia research may shed a different light on the subject.

ANESTHESIA STUDIES

General anesthesia consists of amnesia, hypnosis, analgesia, and areflexia[322]

[242]. Anesthesia studies address the mechanism(s) of hypnosis, or loss of consciousness. Most anesthetic agents decrease global cerebral metabolism [241, 242] and cause a variety of cerebral blood flow (CBF) decrements. Most general anesthetics decrease spontaneous cortical activity, leading to slow oscillations independently of subcortical structures.

Inducers of anesthesia include paradigmatic agents, such as propofol. The anesthetic effects of propofol correlate with the activity of GABA receptors. Propofol causes profound CBF decreases at the medial thalamus, the cuneus, the precuneus, the posterior cingulate and the orbitofrontal cortices. Ketamine is a notable exception. It increases high frequency EEG oscillations, thalamic and limbic metabolic activity, and decreases activity of somatosensory and auditory regions. Ketamine enhances regional cerebral metabolic rate.

The hyperpolarization block of thalamocortical neurons disrupts the circuitry required for consciousness [237]. It interferes with the high frequency rhythms characteristic of conscious states. Intralaminar thalamic nuclei support cortical arousal. The thalamus plays an essential role by seemingly receiving signals from the ascending arousal system as well as by its intimate participation of corticothalamic and thalamocortical circuits [241, 242]. However, it is unclear if the thalamus is the primary site for general anesthetics. Instead, it looks like thalamic effects of anesthesia are secondary to a primary cortical action of anesthetic agents[323]. The decrease in thalamic activity likely occurs secondary to a decrease in afferent excitatory corticothalamic circuits dominated by glutamatergic and GABAergic receptors.

Most general anesthetics suppress the frontoparietal association cortices. The largest decreases in CBF with propofol anesthesia are found in the cuneus, precuneus, posterior cingulate, retrosplenial and frontal cortices [242]. Conscious states would conceivably arise from intermediate level representations modulated by attention and resonating in synchrony [243]. The coactivation of the PCC[324], the dorsal attention network and the frontoparietal control network plays a critical role in allocating attention [173, 244].

General anesthetics disrupt the connectivity of frontoparietal and mediolateral networks. As reviewed by Hudtez [237], cortically evoked responses are not fully blocked under anesthesia, including thalamocortical information transfer. The same does not occur with cross-modal interactions such as those that take place between visual and auditory cortices. There is a loss of higher-level analysis as evidenced by preferential decrease activation of higher-order information-processing regions, but not at primary sensory areas: *"loss of consciousness is not due to a simple block of corticofugal information transfer, but presumably to a lack of higher-order integration in the cortex".*

Reduced PCC and prefrontal connectivity are associated with loss of consciousness. In fact, general anesthetics disrupt connectivity between distant brain regions while preserving local connectivity [245], in support of the GNW theory. A similar effect of anesthetics seems to occur in the context of thalamocortical and corticothalamic connectivity:

> *"...cortical activation to auditory stimuli persisted, confirming that anesthetic unconsciousness cannot be explained by cortical deafferentation or a diminution of cortical sensory activity. Thus, these findings support the theory that the cause of anesthetic unconsciousness is a failure of information integration that appears to correlate with a dysfunction of the nonspecific thalamocortical system".*

Integration[325] among cortical regions, particularly between frontal and parietal, and parietal and temporal areas [245, 246, 318], is clearly decreased during anesthesia. A recent study [247] of coma-causing brainstem lesions showed that a small region in the rostral pontine tegmentum was associated with coma-cause lesions. In healthy adults this brainstem site is functionally connected to the ventral anterior insula and the pregenual anterior cingulate cortex. This finding seems to minimize the role of the thalamus and reinforce the importance of the cortices as the primary regions supporting consciousness.

The anterior insula participates in the salience, limbic, and DMN networks, while the pregenual anterior cingulate cortex is an important part of the DMN. Both of these regions match the distribution of von Economo neurons. These spindle-shaped neurons are found in humans, other great apes, whales, dolphins, and elephants. Their larger size and dendrites across cortical layers integrate and transmit information across large brains and are supposed to participate in some sort of self-awareness.

Dexmedetomidine causes loss of functional connectivity between the DMN and the thalamus as well as bidirectional changes between frontoparietal networks and cerebellar clusters. Dexmedetomidine is a highly selective alpha-2 agonist that decreases the activity of noradrenergic neurons at the locus ceruleus in the brain stem. This medication causes loss of consciousness by disrupting local and global efficiency[326] of large-scale brain networks [248]. It reduces average brain connectivity without changing its distribution.

From an anatomical perspective, networks overlapping the DMN and the nonspecific thalamus are identified as important hubs for consciousness in anesthesia paradigms. From a connectivity perspective, these findings support the

Integrated Information Theory of consciousness.

As previously discussed, this blueprint is consistent with the empirical support of the IIT. I will show next how anesthesia studies help us understand the actions of these integrated networks beyond the IIT approach.

FEEDBACK CONNECTIVITY IN ANESTHESIA

Disruption of feedback connectivity may explain loss of consciousness with general anesthetics [249]. Jordan and colleagues [250] correlated EEG directional connectivity[327] (DC) and fMRI functional connectivity (FC) during propofol anesthesia. Changes in FC in fMRI correlated with simultaneous changes in DC in the EEG. This finding indicates that the mechanism underlying anesthesia-induced unconsciousness consists in the decoupling of "higher-order" frontal and "lower-order" parietal processes, *i.e.* a breakdown of directional feedback connectivity.

It was shown [251, 252] in human subjects that feedback[328] and feedforward frontoparietal connectivity decrease precipitously with propofol anesthesia. However, feedforward connectivity returned to baseline while individuals were still unconscious. The recovery of feedback connectivity was verified with the return of consciousness.

Reduction in cortical feedback activity seems to facilitate and precede sleep and the unconsciousness provided by propofol anesthesia [253]. GABA receptor agonists such as propofol inhibit directional connectivity from anterior to posterior regions.

Lee and colleagues [254] assessed directional connectivity across the frontal, parietal, and temporal regions of surgical patients using normalized symbolic transfer entropy in EEG measurements. They found that propofol (mostly an intravenous GABA agonist), sevoflurate (a volatile GABA agonist with diverse molecular effects), ketamine (working through N-methyl-D-aspartate NMDA receptors without inhibition of the thalamus) disrupt anterior to posterior directional connectivity despite belonging to molecularly-distinct classes of anesthetics.

> *"Most of the data currently available suggest that general anesthesia is correlated with a decrease in top-down connectivity from frontal to parietal, temporal, and occipital cortices as well as from parietal to occipital cortices" [254].*

EFFECTS OF ANESTHETICS ON PYRAMIDAL NEURONS APICAL DENDRITES

Gamma-amino butyric acid (GABA) and N-methyl-D-aspartate (NMDA) receptors are the most important molecular targets of general anesthetics. In their review, Uhrig and colleagues [241] point out that anesthesia occurs through the enhancement of inhibitory currents *via* the GABA receptor pathway and glycine protein channels, through the decrease of excitatory currents potentiated by glutamate and acetylcholine, and through the enhancement of background potassium leakage currents.

$GABA_A$ receptors represent the main inhibitory brain receptors [248]. $GABA_A$ receptors are ionotropic, while $GABA_B$ receptors are metabotropic. The binding of GABA to $GABA_A$ receptors opens a Cl^- channel, which causes membrane hyperpolarization leading to a reduction in excitatory signals. Glycine, a less common inhibitory transmitter in the brain, also activates ionotropic receptors that allow the influx of Cl^- ions[329].

Glutamate is the main excitatory transmitter of the brain, acting on multiple metabotropic (G protein coupled) and ionotropic (ligand-gated ion channels) receptors. Metabotropic receptors indirectly gate ion channels through the action of second messengers. There are three ionotropic glutamate receptors, named after the synthetic agonists that stimulate them: NMDA, AMPA and kainate. NMDA[326] receptor stimulation requires the co-activation of glutamate with either glycine or D-serine. They produce an influx of positive ions, such as Ca^{2+}, K^+, and Na^+, and generate depolarizing currents.

Barbiturate and volatile anesthetics stimulate $GABA_A$ and glycine receptors and inhibit NMDA and AMPA glutamatergic receptors in various degrees. Some volatile anesthetics also stimulate kainate receptors. Kainate receptors may have postsynaptic excitatory effects or presynaptic inhibitory effects through the modulation of GABA release. Ketamine inhibits NMDA-mediated glutamatergic transmission and produces a limited stimulation of $GABA_A$ receptors.

These receptors are distributed throughout the cortex, thalamus, striatum and brainstem [323]. General anesthetics slow cortical oscillations with the exception of ketamine. Ketamine enhances high frequency cortical oscillations. Anesthetic mediated oscillatory cortical effects precede thalamic changes[331], which is consistent with a primary role for the suppression of corticothalamic feedback to the thalamus.

General anesthetics decrease top-down connectivity of sensory pathways. In humans top-down processing predominates over bottom-up processing. Loss of neural synchronization between anterior and posterior brain regions appears to represent this disruption, which seems not to vary regardless of anesthetic agent. Meyer's [255] comprehensive review proposes a shared mechanism for these effects associated with the action of general anesthetic on apical dendrites of pyramidal neurons.

As previously reviewed, pyramidal neurons integrate top-down and bottom-up brain signaling. Corticocortical top-down (feedback) signaling is projected to distal apical dendrites. These signals are carried towards the cell body by NMDA receptors and Ca^{2+} potentials. According to Meyer [255], the initial input is carried by NMDA spikes. Ca^{2+} potentials closer to the pyramidal cell body relay these signals to the cell body itself. As previously reviewed in the chapter addressing cortical signals of consciousness, this calcium spike produces more action potentials than direct suprathreshold stimulation of the pyramidal cell body[332]:

> *"Crucially, several lines of evidence suggest that general anesthetics can interfere with dendritic signal conduction and with the coupling mechanism between dendritic and somatic action potential initiation zones in pyramidal neurons."* (Table **6** and Fig. **14**).

Table 6. General anesthetics effect on pyramidal cells dendritic signaling. Coincident somatic and dendritic stimulation convey the top-down / bottom-up realization of information [255].

GENERAL ANESTHETICS TOP-DOWN / BOTTOM-UP REALIZATION OF INFORMATION DISRUPTION	
Suppression of the generation of dendritic calcium action potentials (generated at the main apical dendrite bifurcations) carrying signals along the main apical dendrite towards the soma	Urethane, pentobarbital
Suppression of NMDA potentials carrying signals from peripheral tuft dendrites toward the calcium potential initiation zone	Ketamine, xenon, nitrous oxide Sevoflurane, desflurane (lesser degree)
Up-regulation of inhibitory neurons (in layers 2 and 3) that suppress the generation of dendritic calcium action potentials	Inhibitory effect of GABA[333] stimulation
Inhibition of the hyperpolarization-activated current I_h[334] that causes uncoupling of pyramidal cells' somatic and dendritic compartments	Several volatile and intravenous anesthetics, such as halothane, enflurane, isoflurane, propofol, pentobarbital, ketamine

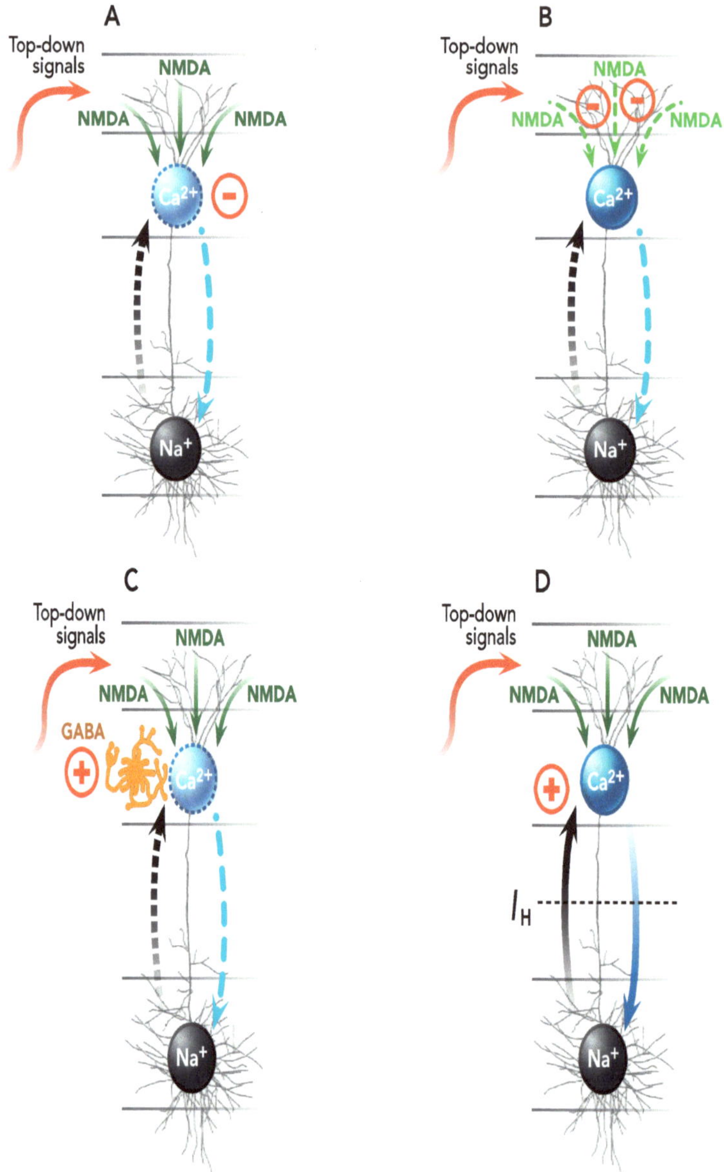

Fig. (14). Disruption of pyramidal cells dendrite's signaling. The figure illustrates four mechanisms of general anesthetics disruption of top-down/bottom-up realization of information: A) Suppression of calcium action potentials, B) Suppression of NMDA potentials, C) Upregulation of inhibitory neurons, D) Inhibition of the hyperpolarization-activated current I_h (Reproduced with permission from Wolters Kluwer, Inc.).

In summary, top-down processes play an essential role in conscious perception. General anesthetics seem to act by disrupting top-down signaling and possibly by disabling mechanisms matching top-down and bottom-up informational streams at cortical pyramidal cells. These effects likely underlie the disruption of gamma synchrony across different brain regions secondary to general anesthetics.

This hypothesis [255] does *"not claim that an inhibition of dendritic signaling in cortical pyramidal neurons constitute a unifying mechanism that, alone, would explain how each and every general anesthetic cause unconsciousness"* as there is evidence of subcortical anesthetic effects For example, it does not include the anesthetic effects of opiates. However, it critically advances the understanding of the physiological context of ketamine anesthetic actions. Ketamine impairs top-down connectivity. It increases, rather than decreases thalamic metabolism.

Anesthesia studies in consciousness support the main hypotheses of this blueprint, namely the importance of top-down agency in conscious experiences and the realization of information through the strange loop

'expectation.information'expectation.information
'top-down.bottom-up'top-down.bottom-um

QUANTUM OPERATIONS IN DENDRITES

According to the Orch OR theory [43], consciousness emerges from dendritic-somatic integration of information. Dendritic-somatic membranes generate local field potentials that give rise to the EEG, including coherent gamma synchrony. The theory proposes that processes in microtubules mediate consciousness and regulate neuronal firing. These processes are thought to relate to conformation transitions that give origin to collective oscillations in tubulin [256]. There is some evidence to suggest that anesthetics dock to tubulin and produce a change on their collective oscillations.

General anesthetics can alter pyramidal cell dendritic signaling by several other mechanisms, as reviewed in Table **6**. In addition, inhalational anesthetics disrupt protein-protein interactions at the PDZ domain[335], which are necessary for the assembly of synaptic multiprotein signaling complexes [257]. Inhalational anesthetics are able to dose-dependently and specifically disrupt the protein interaction between the synaptic PDZ domain and NMDA receptors or the neuronal nitric-oxide synthase. A recent report [258] showed that propofol and etomidate (a $GABA_A$ anesthetic) restrict syntaxin1A mobility on the plasma membrane. Syntaxin1A is a key membrane protein part of the neurotransmitter release machinery. Syntaxin1A-mediated neurotransmission is a highly conserved evolutionary mechanism that underlies exocytosis in every neuron. This mechanism can dramatically change temporal brain dynamics and cause loss of functional connectivity.

These considerations do not invalidate the possible existence of anesthetic effects on the neuronal microtubule cytoskeleton. However, they indicate that cytoskeleton paradigms have not been at the forefront of theories about the mode of action of general anesthetics.

Nonetheless, the Orch OR theory and the Integrated Information Theory fail in similar ways. The bottom-up approach of both theories is inconsistent with evidence from the study of anesthesia suggesting a crucial role for top-down signaling in human consciousness. They share a cryptic intelligent bias and do not recognize the role of the subject as an operator of conscious experiences.

Hameroff and Penrose [43] and Craddock and colleagues [256] propose that quantum phenomena in the oscillatory activity of tubulin underlie local field potentials and play a causal role in synaptic firing regulation and gamma synchrony. Even if this is true, conscious experiences are not built up from bottom-up processes, but through the interaction between top-down and bottom-up circuits.

These researchers have asked the wrong question. In light of the evidence, a more adequate question would have been: considering that the primacy of top-down signaling in consciousness has conceptual and empirical support, can oscillations in tubulin respond to top-down activity?

If quantum operations truly happen in the brain, this could render neuronal processes an environment for decoherence marked by top-down systems expectations. Orch OR makes objective reduction the axiomatic justification of consciousness. Instead, objective reduction is seen here as the final consequence of top-down processes originating from the subject actions.

CONSCIOUSNESS WITHOUT A CORTEX

From an anatomical perspective, damage to several areas of the human brain may cause cognitive deficits and other impairments without an effect on consciousness. Morsella and co-workers compiled a list of regions in the central nervous system that, if extirpated, lead to deficits, but not to the eradication of consciousness [259]. The elimination of the spinal cord, cerebellum, amygdalae, hippocampi, cortices including the prefrontal cortices, and even an entire hemisphere do not render the human brain incapable of generating consciousness. Several of these regions elaborate on conscious experiences but should not be considered the "seat" of consciousness. Instead, the authors pointed towards a mesencephalic[336] seat of consciousness[337] [259].

The GNW theory describes the processes that underlie consciousness in the intact human brain, and their disruption in the unconscious or minimally conscious human brain. However, consciousness may still emerge in the absence of a cortex. This may be true for some children born without the cerebral cortex.

Although most anencephalic[338] children die shortly after birth, some survive for longer periods. They have hydranencephaly, where cerebrospinal fluid replaces lost cerebral tissue. Based on their clinical responsiveness, children born without most of the cortex can be considered conscious [258]. Merker closely examined the behaviors of five anencephalic children between the ages of ten months and five years. These children were awake, often alert and responsive to their environment[339]. Anencephalic children smile and laugh[340]. They may show behavioral initiatives and instrumental behaviors, such as kicking trinkets for noise or activating switches of their favorite toys [258]. Those behaviors are evidence of associative learning of actions and their consequences.

These children frequently suffer from absence seizures. Absence seizures, historically known as petit mal, are generalized seizures that typically begin and end suddenly. Their only characteristic symptom is the sudden loss of consciousness, without any specific motor activities. Patients typically stop all their activity for a few seconds. They look like they are staring off into space with a blank look. Parents can easily distinguish episodes of loss of consciousness due to absence seizures in their children with anencephaly. They can tell if their children are aware, or not, of their surroundings.

In summary, these children displayed behaviors consistent with general consciousness[341] and access consciousness. Of course, the author could not probe into the contents of their specific consciousness, but some of their behaviors seemed to reflect phenomenal consciousness.

THE SPRAGUE EFFECT

Unilateral removal of the posterior cortical visual area in cats causes a contralateral loss of response to visual stimuli. Removal of the contralateral superior colliculus or cutting the connection between colliculi reverses the loss, in what is known as The Sprague effect[342], first reported in 1966. The disruption of cortical fibers that project to the superior colliculi unbalances the mutual inhibition of colliculi. Their disruption disinhibits the contralateral colliculus and releases its inhibitory effects upon the ipsilateral colliculus[343]. Cutting down their mutual inhibition restores the suppressed colliculus, which is located on the same side of the cortical lesion. Inhibitory cortical fibers cross over to the contralateral colliculus.

The Sprague effect illustrates the limitations built into the use of descriptive brain landscapes in theories of consciousness: in the functionally intact human brain, consciousness seems to correspond to activity in the global workspace; in the damaged or cortically deactivated human brain, secondary effects in the midbrain may contribute to loss of consciousness.

THE BRAIN STEM CONVERGENCE SUPPORTS A PHENOMENAL SUBJECT-CENTRIC MOTOR ANALOGIC SIMULATION

The superior colliculus (SC), located in the midbrain roof (tectum), is the only region where *"senses are topographically superposed in laminar fashion within a common, premotor framework for multiple effector control of orienting"* [261]. Any sensory modality requiring phasic orientation must have a representation in the SC: vision, audition, somatosensory, pain, infrared, electroreceptive, magnetic, and even echolocation systems.

The SC receives cortical afferents from cortical layer 5. In primates, the SC receives parietal and temporal information representing dorsal ("where it is") and ventral ("what it is") streams[344] of the visual system. Connections with the cortex are monosynaptic, with no more links than corticocortical and corticothalamic connections. The SC may play a causal role in actionable selection of target [261]. SC lesions do not necessarily prevent target acquisition, but cause loss of scope and sophistication in target selection. The integrative SC function is *"anatomically subcortical and functionally supra-cortical"* [260].

Fabbro and colleagues reviewed behavioral and neuronal evidence of consciousness in animals[345], including fish, amphibians and mammals [262]. For example, according to the authors, several species of fish have long term memories, change their social and reproductive behavior after observing other fish interactions, assess their own status and behave based on the assessment of third-party behaviors.

Several vertebrates represent spatial aspects of the world, the self, and time. Their behaviors provide indirect evidence of phenomenal experiences emerging early in evolutionary history. The authors attributed phenomenal consciousness to the multimodal capability to integrate sensory stimuli and generate multisensory representations of body and environment. In this sense, subcortical systems made of reciprocal connections between brain stem and central thalamic nuclei with a phylogenetically[346] old cortex exist in all vertebrates and allow for reentrant interactions, seen as the origin of a primary consciousness[347]. They considered that all vertebrates most likely have some sort of sensory-phenomenal, affective consciousness controlled by circadian cycles and various states of being[348] [262].

More specifically, these authors indicated that multisensory integration happens at the roof of the midbrain, in the SC or the optic tectum, where the brain forms a topographic map representing the surrounding world. These subcortical structures appeared early in evolutionary history.

INTEGRATION FOR ACTION

"Higher" from a cognitive perspective does not necessarily mean "higher" from a control perspective. The mesodienchephalon[349] houses the *"major integrative structures of broad functional scope shared by all vertebrates"* [258].

The primary connections of the optic tract span the region between the midbrain and the diencephalon, including the pretectum[350]. The region between the hypothalamus at the bottom of the diencephalon and the SC at the roof of the midbrain integrates goal-directed adaptive motivated behaviors. This multimodal integration coordinates the movement of the eyes, head, and body with motor navigation. The intermodal integration stabilizes the perception of the world during movement to the benefit of adaptive mobility[351].

Merker's vision-as-action paradigm is compatible with an action-oriented view of consciousness in which action selection depends on target selection and motivational state.

The SC provides a coordinate system for the laminar superposition of spatial senses in a pre-motor framework. This system materializes the analogic integration of a mobile body with a movement-stabilized spatial perceptual environment in vertebrates [260]. Moreover, periaqueductal gray connections to deep collicular layers can provide information for emotion-related behaviors, such as *"defensive, aggressive, vocal, sexual, and pain-related ones[36]"*:

> *"...the [original] 'world' of the proposed neural reality simulator presumably amounted to no more than a two-dimensional screen-like map of spatial directions on which potential targets might appear as mere loci of motion in an otherwise featureless noise field, defined more by their displacement than by any object features."*

The upper brainstem provides an oculocentric, perspective-based coordinate system that separates objects of consciousness from consciousness itself. The interface *"possesses the essential attribute of [phenomenal, primary] consciousness"* implicit in its fundamental role of integration for action. But what makes it actually conscious?

THE UPPER BRAINSTEM MAY SUPPORT PREDICTIVE CODING

The upper brainstem fits a spatially mobile body with a movement-stabilized spatial environment through shared coordinate system integration. Information is encoded for global sequential actions of the organism acting towards represented objects at a descriptive level consistent with phenomenal experiences.

Other than the olfactory system, the superior colliculus is the only place outside the cortex-thalamus unity with fast oscillations in the gamma range that parallel this cortical activity [260,262]. The brain generates gamma oscillations locally [263]. The network serves critical adaptive behaviors selecting the most important information and leading vision-based behavior away from distractive stimuli[352] [264, 265].

Corticotectal SC activity is intrinsically coupled, and functionally interacts across multiple temporal and spatial scales, including the amplitude of delta, spindle, low gamma, and high-frequency oscillations (>100Hz). It is also temporally coordinated by the phase of slow cortical oscillations[354]. Up states of cortical slow oscillations strongly entrain on downstream subcortical networks. The spindle-like activity in the SC displays a *"strikingly similar cross-frequency phase-amplitude relationship as thalamically generated spindles"* [266].

EVIDENCE OF CONSCIOUSNESS IN ANENCEPHALIC CHILDREN CAN MAKE THE GNW A FALSIFIABLE THEORY FOR THE HARD PROBLEM OF CONSCIOUSNESS

These findings suggest that the upper brainstem makes inferences that are consistent with the principles of Predictive Coding as well as with the statistical physics represented by the ensemble density $q(\vartheta,\lambda)$, *i.e.* the probability density that causes ϑ would be selected from an ensemble of environments. Moreover, sustained oscillatory activity allows for the existence of activity states that are compatible with the states of sustained phenomenal consciousness.

A generational model, with its conditional expectations, is a simulation in itself. Phenomenal experiences may conceivably emerge with the onset of sustained oscillatory rhythms depicting the sustained expectations in an integrative action-based model. Integration at upper brain regions becomes the simulation.

Traces of phenomenal consciousness may thus emerge with the onset of rhythms that sustain an organism's internal expectations [196,266-268]. A generational model is an implicit simulation paradigm that is consistent with clinical observations of anencephalic children. These observations challenge the GNW model as an explanation for the Hard Problem of Consciousness.

The laminar superposition of spatial senses in a pre-motor framework at the upper brain stem allows for the multimodal integration that leads to action. The interface presents the essential attributes supporting phenomenal consciousness. The superior colliculus shows gamma range oscillation that parallels cortical activity. The superior colliculus may support sustained rhythms encoding Predictive Coding expectations, and therefore may transform the upper brain stem integration into an analogic self-simulation required for subjective phenomenal experiences.

The descending projections of cortical layer 5 are considered to convey the final summary of cortical operations to the rest of the brain [269]. Layer 5 *"efferent cascade"* represents the *"completed record of cortical labor"*: it reflects the outcome of the conjoint cortical feedback and feedforward projections across cortical layers, a condition that seems necessary for information enter consciousness.

Throughout this book I stressed the difference between a description and a theory, such as the difference between natural history and evolutionary theory. GNW theory, as a theory for the *cause* of consciousness, can be made falsifiable with evidence of *"consciousness without a cortex"*.

CONCLUSIONS

General anesthetics seem to act by disrupting top-down signaling and disabling mechanisms matching top-down and bottom-up informational streams on the apical dendrites of cortical pyramidal cells. This is not the sole explanation for all the effects of general anesthetics, but an effect upon brain signaling that is shared by several general anesthetics.

Orch OR theory proponents hypothesize that quantum phenomena underlie oscillatory phenomena in dendritic tubulin. The Orch OR model rests on a causal origin of conscious experiences from the bottom-up. Instead, general anesthetics studies suggest a crucial role for top-down projections and the matching of top-down and bottom-up streams [255].

Nonetheless, top-down effects on quantum phenomena may render neuronal processes an environment for decoherence marked by expectations of the top-down system expectations. Orch OR makes objective reduction the axiomatic justification of consciousness. Instead, objective reduction is seen here as the final consequence of top-down processes that originate from the actions of the subject.

Inference procedures stemming from a core system of self-awareness, which overlap with the Default Mode Network and networks for social perception likely play an important role in endogenous conscious experiences. The DMN anticorrelates with Task-Positive networks. Neither Task-Positive networks nor DMN computations directly generate a subject. The subject more likely emerges noncomputationally, and secondary to the realization of information by networks that run resting states and external tasks. Conversely, the activity of these networks identifies what the subject does. This view supports an action-centered origin of consciousness.

The GNW theory identifies the thalamus as part of a core of consciousness. The thalamus is not a core part of the DMN. Anesthesia studies suggest that thalamic changes are secondary to primary cortical actions. Theories that focus on the DMN as well as the GNW theory can be reconciled through the phenomenological descriptions provided by classic Zen texts of a field of consciousness that is not bound to contents.

Children born without a cortex can display behaviors associated with consciousness. They can smile and laugh, show initiative and instrumental behaviors. Their parents can easily distinguish episodes of loss of consciousness due to absence seizures. In the damaged or cortically deactivated brain, secondary effects in the midbrain contribute to the loss of consciousness. The upper brain stem may support sustained rhythms encoding Predictive Coding expectations. This subcortical region provides an interface with the essential attributes of phenomenal consciousness.

Phenomenal experiences may conceivably emerge with the onset of sustained oscillatory rhythms depicting sustained expectations in an integrative action-based model. These observations are consistent with the blueprint hereby presented.

NOTES

[317] In a simplified hierarchy, Carhart-Harris and Friston placed the DMN at the top of the brain systems, the salience and dorsal attention at intermediate levels, and the thalamic and unimodal sensory cortices at a lower level [222]. Each level supposedly suppresses free-energy from its subordinates by improving predictions and reducing prediction errors.

[318] The claustrum may also play an important role in the corticocortical interactions necessary for consciousness. The claustrum has strong functional connections with the insula and cingulate cortex. It is considered to mediate multisensory integration, salience detection, and attention. The claustrum is a thin and irregular sheet of neurons located under the neocortex and connecting with several cortical regions. In rodents [331], deep isoflurane anesthesia attenuated the claustrum functional connections, but not necessarily the connections of the rest of the cortex. The main affected functional connections were those with the medial prefrontal cortex and the mediodorsal thalamus. This study supported the idea of the claustrum coordinating cortical areas processing sensorimotor information relevant in bilaterally coordinate behaviors, such as eye movement or

exploratory whisking.

[319] *"Hongzhi's expression of silent illumination is the culmination of a sophisticated Zen Buddhist teaching tradition inspired by the founders of the Chinese Caodong lineage (pronounced 'tsow-dong'), later called Sōtō in Japan. This Chinese Sōtō tradition developed a theoretical understanding through the dialect of insight into universal truth and its interplay with the particulars of the phenomenal world. His work was later elaborated and developed in Japan by Zen Master Dōgen and the Sōtō tradition that derived from him"* [238].

[320] *"The field of boundless emptiness is what exists from the very beginning. You must purify, cure, grind down, or brush away all the tendencies you have fabricated into apparent habits. Then you can reside in the clear circle of brightness. Utter emptiness has no image; upright independence does not rely on anything. Just expand and illuminate the original truth unconcerned by external conditions. Accordingly we are told to realize that not a single thing exists. In this field birth and death do not appear. The deep source, transparent down to the bottom, can radiantly shine and can respond unencumbered to each speck of dust without becoming its partner. The subtlety of seeing and hearing transcends mere colors and sounds. The whole affair functions without leaving traces, and mirrors without obscurations. Very naturally mind and dharmas emerge and fulfills the way of non-mind. Enacting and fulfilling the way of the non-mind, finally you can rest. Proceeding you are able to guide the assembly. With thoughts clear, sitting silently wander in the center of the circle of wonder. This is how you must penetrate and study"* [238].

[321] In support of this view, in sleep and sedation thalamocortical connectivity of nonspecific (intralaminar) nuclei is preferentially reduced, in comparison to specific thalamic systems [229, 316, 317]. It is worthwhile noting that specific thalamocortical fibers project to cortical layer 4 whereas nonspecific fibers terminate primarily in superficial layer 1, which is the main afferent layer for top-down projections [255]: *"...while specific thalamocortical projection would carry the content of sensory signals (e.g.* auditory or visual information) to the cortex, the nonspecific projection would be require to conjoin these fractured perceptual components into unified cognitive events by synchronizing the individual cortical firing patterns. This conjunction process would be implemented by the summation of specific (to layer 4) and nonspecific (to layer 1) thalamic signals along the radial dendritic axis of cortical pyramidal neurons". These considerations may provide physiological credence to Master Hongzhi phenomenological description of a subtle bright field of consciousness. They may be part of the discussion addressing Fringe consciousness and the phenomenology of the cognitive access to perceptual experience [319 - 322]. It is worthwhile to note the parallel between specific/nonspecific distribution of the tuft/mitral cells with specific/nonspecific thalamic projections. This relationship may support the binding of information across cortical regions.

[322] I should refer the reader to the excellent reviews of Cerebral Mechanisms of General Anesthesia by Uhrig, Dehaene and Jarraya [241] and Consciousness and Anesthesia by George Mashour [242].

[323] The thalamus seems to participate of the unconsciousness of anesthesia through a hypersynchrony of the alpha band with the frontal cortex. Restoration of consciousness is

associated with increased thalamic firing above baseline activity, consistent with a thalamic switch "rebooting" thalamocortical circuits.

[324] This is particularly relevant for those authors that do not see the possibility of consciousness without attention: *"these hypotheses and predictions about the attentional functions of the PCC basically state that dorsal parts of the PCC are responsible for external attention to the environment and ventral parts of the PCC are responsible for internal attention"* [173].

[325] Functional correlation does not necessarily imply effective communication. The high correlation of hypersynchrony may cause loss of discriminable brain states while reduced connectivity may decrease information integration.

[326] Network efficiency is a graph theoretical measure of information exchange.

[327] To quantify DC they applied *"symbolic transfer entropy (STEn), which distinguishes driving and responding subsystems in complex nonlinear dynamics and detects asymmetry in their mutual interaction"*. Directional connectivity was inferred when *"electroencephalographic symbolic transfer entropy was a applied as a nonlinear measure of mutual interdependencies between underlying physiological processes. The relationship between FC of resting-state networks of the brain (z values) and DC was analyzed by a partial correlation"*.

[328] Of note, reduction of feedback activity in rats is also associated with anesthetic unconsciousness, although in these animals a balance between feedforward and feedback connectivity characterizes wakefulness. In contrast, conscious humans show a dominance of feedback connectivity.

[329] The unitary Cl^- inflow of glycine-gated channels is slightly larger than $GABA_A$-gated channels because the diameter of the glycine-gated pore is slightly larger than the $GABA_A$ pore [323].

[330] NMDA receptors are unique among ionotropic receptors because their opening is both voltage- and neurotransmitter-dependent. The mechanism is different from other voltage gating channels that suffer conformational changes due to membrane voltage changes. In NMDA receptors, depolarization removes a specific Mg^+ ion plug that blocks the current allowing Ca^{2+} and Na^+ to enter. The drug of addiction phencyclidine (PCP) binds to a different channel site and blocks the NMDA receptor [323].

[331] In rodents electrophysiological changes in midline thalamus precede neocortical changes in both sleep and anesthesia [255]. The conflictive results with human encephalography may be due to the asymmetrical top-down/bottom-up signaling in humans.

[332] *"The two spike initiation zones (the one in the apical dendrite and the 'conventional' one at the axon hillock) mutually influence one another: a single action potential triggered at the soma can back-propagate into the dendrite and lower the threshold for calcium potentials by half. Due to this mechanism, cortical pyramidal cells are in a unique position to associate top-down signals arriving at their dendrites with bottom-up information arriving at their soma, and dendritic calcium potentials, along with the subsequent characteristic action potential burst at the soma, may represent a distinctive signature of coincident somatic and dendritic activation"* [255].

[333] *"Specifically, a short-lasting dendritic inhibition component (up to 150ms) is mediated by $GABA_A$ receptors, whereas a long-lasting component (up to 400ms) relies on*

$GABA_B$ receptors."

[334]I_h is a leak conductance that attenuates the propagation of synaptic potentials under physiological conditions: *"When I_h is blocked, somatic activation alone may trigger a dendritic calcium spike and the subsequent burst of somatic action potentials, leading to the breakdown of the unique associative mechanism afforded by pyramidal cells between input to their somatic and dendritic compartments"* [255].

[335] The name PDZ derives from the first three proteins in which these domains were studied: PSD-96/SAP90, Dlg, ZO-1). The effect occurs at hydrophobic sites on the target protein the concentrations equivalent to anesthetic potency [257].

[336] The mesencephalon [324] is a structure apparent at the early stages of the embryonic neural tube development (proencephalon, mesencephalon, rhombencephalon, and the caudal neural tube). The telencephalon gives origin to the telencephalon and the diencephalon, while the mesencephalon develops into the midbrain. The midbrain is the smallest part of the brain stem. It connects crucial components of the motor system (particularly the cerebellum, basal ganglia, and the cerebral hemispheres). The midbrain contains parts of the auditory and visual systems, and project connections to the extraocular muscles controlling eye movements [221]. The midbrain is the original place for the interface of visual and auditory information into global motor behaviors.

[337] Interestingly, they also stressed the importance of the olfactory system as a gateway to understand consciousness, and alluded to patients who had bilateral olfactory bulbectomies but still had olfactory memories. Bulbectomy is the extirpation of the olfactory bulbs. Their observation is consistent with the idea of "odor-objects" being organized at the primary olfactory cortex, rather than at the olfactory bulbs.

[338] Anencephaly: lack of development of the cerebral hemispheres.

[339] *"These children are not only awake and often alert, but show responsiveness to their surroundings in the form of emotional or orienting reactions to environmental events, most readily to sounds but also to salient visual stimuli (optic nerve status varies widely in hydranencephaly). They express pleasure by smiling and laughter, and aversion by "fussing", arching of the back and crying (with many gradations), their faces being animated by these emotional states. A familiar adult can employ this responsiveness to build up play sequences predictably progressing from smiling, over giggling to laughter and great excitement on the part of the child. The children respond differently to the voice and initiatives of familiars, and show preferences for certain situations and stimuli over others, such as a specific familiar toy, tune or video program, and apparently can even come to expect their regular presence in the course of recurrent daily routine."*

[340] They smile and laugh, "*in the specifically human manner, which is different from that of our closest relatives among the apes*". According to Merker, these characteristic human manifestations emerge from the human brainstem.

[341] General consciousness: what makes a state conscious at all (it "roughly" corresponds to the level of consciousness). Specific consciousness: what gives a state its specific phenomenal character (it "roughly" corresponds to the content of consciousness). Access consciousness: mental states which content can be directly used in actions [163]. Phenomenal consciousness: experience of "something it's like" [320]. *"Someone possesses phenomenal consciousness if there is any type of subjective experiences that is currently presented for him/her"* [325].

[342] In Sprague's own words [326]: *"So clearly, we discovered, there is marked interaction between cortical and midbrain levels in visual orienting. Only when the cortical lesions were much larger and included all of what is now recognized as visual sensory and association cortex, could the deficit be termed hemianopia with total and enduring loss of orienting. However, an astonishing recovery of orienting followed subsequent ablation of the colliculus on the opposite side of the brain. This dramatic result, later called the Sprague effect, forced a new concept of brain function: the paired orienting centers in the superior colliculi have a reciprocal inhibitory mechanism that suppresses activity in one colliculus when the other is directing orienting responses toward contralateral space. This inhibitory mechanism is unbalanced by the effects of the cortical lesion, so that the colliculus ipsilateral to the lesion receives a chronic influx of inhibition, resulting in lack of orienting contralaterally and the presence of visual neglect of contralateral space. This idea was supported by cutting the collicular commissure, which also restored orienting."*

[343] After cortical damage, indirect inhibitory projection pathways crossing midline in the collicular commissure disrupt the balance between the colliculi [260]. In addition to recovery of orientation, Merker emphasizes the partial restoration of vision. He pointed to a similar phenomenon, which occurs after inactivation of the junction of the occipital, parietal and temporal lobes in cats, and causes contralateral neglect of the space outside. Cooling the contralateral superior colliculus partly reverses the effect. Weddel reported a case of Sprague effect in a human patient [327]. His patient had a midbrain tumor and developed left side neglect due to a right side frontal lobe injury. Bilateral orientation returned as the tumor progressed and caused damage to the left superior colliculus, contralateral to the frontal cortical lesion. The temporary restoration of bilateral orientation was lost when the tumor caused additional damage to the right superior colliculus.

[344] The full distinction between the roles of dorsal and ventral information stream is controversial. I will avoid reviewing the arguments. The emphasis here is on the convergence of information at the SC.

[345] Their classification of consciousness followed Damasio's [105,108] classification:1. Proto-self: interconnected and temporarily coherent collection of neural patterns representing the state of the organism, moment by moment, at multiple levels of the brain. A template for the organization and coordination of all sensations and primordial emotions (we are not conscious of the proto-self); 2. Core self: the transient protagonist of consciousness, generated by any object that provokes the core-consciousness mechanism. It is considered a second-order map of the proto-self, modified. Due to the permanent availability of objects, it is continuously generated and appears constant in time. It is a pre-reflective or tacit consciousness of oneself as an immediate embodied subject, including sense of ownership and agency (we are conscious of the core self); 3. Autobiographical self: autobiographical memory constituted by implicit memories of multiple instances of individual experience of the past and anticipated future. It is based on permanent but dispositional records of core-self experiences, which can be activated as neural patterns and turned into explicit images. Each memory operates as "something-to-be-known" and generates its own pulse of core consciousness.

[346] Phylogenetic: the study of the evolutionary history.

[347] Fabbro and colleagues [262] criteria for anatomic and physiologic evidence of primary consciousness: 1) A pattern of EEG activity in the range of 20-70Hz (typically associated to wakefulness and REM sleep); 2) Thalamocortical activity; 3) Widespread brain activation during the processing of sensorial stimuli; 4) Selective synchronization at cortical and brainstem levels of dynamically formed neural networks involved in binding sensory stimuli; 5) Evidence of egocentric maps for the localization of an individual in a given space.

[348] Interestingly, brain-machine interface (BMI) experiments show that nonhuman primates "*can voluntarily opt to dissociate the production of brain activity containing voluntary motor intentions from the translation of these intentions into true muscle actions.*" [207] BMI experiments indicate phenomenal consciousness and incorporation of objects into the simulated body schema in these animals. This line of research provides experimental evidence of animal consciousness.

[349] Mesodienchephalon or the "optic brain" is considered by Merker a more evolutionary valid designation as the historical grouping of the diencephalon with the telencephalon does not follows their actual phylogenetic background [258]. As reviewed, the telencephalon gives origin to the telencephalon and the diencephalon, while the mesencephalon develops into the midbrain.

[350] The pretectum is considered the transitional synencephalum.

[351] "*If, therefore, integration is for action, as proposed here for the mesodiencephalic control system, information-theory poses no obstacle to having an expansive neocortex make its contribution in this regard by convergent projections on the highly conserved and pre-existing machinery of the midbrain and basal diencephalon, which therefore could retain its old integrative functions. Indeed, a bottleneck of this kind is exactly what is necessary to convert the massive parallel and distributed information capacity of the cerebral hemispheres into a limited-capacity, sequential mode of operation featured in action selection for coherent behavior*" [260].

[352] "*After recovery, decorticate rats show no gross abnormalities in behavior that would allow a casual observer to identify them as impaired in the basis of cues in posture, movement or appearance. They stand, rear, climb, hang from bars and sleep with normal postures. They groom, play, swim, eat, and defend themselves in ways that differ in some details from those of intact animals...*" "*Neonatally decorticated rats as adults show the essentials of maternal behavior which, though deficient in some respects, allows them to raise pups to maturity. Some, but not all, aspects of skilled movements survive decortification and decorticate rats performs as readily as control on a number of learning tests. Much of what is observed in rats (including mating and maternal behavior) is also true of cats with cortical removal at infancy: they move purposefully, orient themselves to their surroundings by vision and touch (as do the rodents), and are capable of solving a visual discrimination test in a T-maze...*" "*...Coherent and well-organized behavior is elicited by local stimulation in the mesodiencephalic region of intact animals*" [260].

[353] Event-related LFPs in the SC primarily reflect the incoming visual signal, rather than the motor signal for the ocular saccade. The deactivation of the dorsolateral prefrontal cortex (DLPFC) in monkeys does not affect the LFP activity in the SC associated with stimulus onset. However, DLPFC deactivation decreased high beta (20–30 Hz) and high

gamma (60–150 Hz) power period and the spike rate during the preparatory period for saccades, indicating that top-down control of the SC by the DLPFC may be mediated by beta oscillations [328].

[354] Slow cortical oscillations are generated locally within cortical layer 5 and propagate to other granular and supragranular layers. As neurons in layer 5 are strongly activated during up states of slow oscillations, their downstream target neurons are active preferentially during up states of slow oscillations.

CHAPTER 19

A Problem of Records

Abstract: The problem of records consists of matching neural signals with specific quale. The question does not contemplate why human brains generate subjective experiences, but the archives of the rules that match subjective experiences to objective brain states. Record keeping is a specific non-metaphysical aspect of the Hard Problem that is not algorithmically compressible and therefore non-solvable from a computational perspective. The answer to this question lies in the physics of the universe. Record keeping remains a difficult problem in quantum mechanics. Decoherence can be seen as the loss of information from a system into the environment. It is not clear which property of the environment carries the information from the decohered system. The development of a quantum explanation requires the identification of the property that makes the wave state respond to living things with subjectivity, and not with physical information such as position and momentum. Conscious experiences emerge from spatiotemporal patterns of integrated activity at levels that quantum theory cannot address. Processing of information across hierarchies is consistent with the generation of abstract concepts transcending their conditions of origin. Emergent properties that allow for conscious representation of objects require stable resonance of neuronal activity much beyond the levels of organization addressed by quantum physics. Quantum theories must reconcile recent neuroscience findings that emphasize the role of top-down information in human consciousness. However, strange loops may provide a mechanism for the generation of awareness, but they do not explain the subject who is the experiencer. The relational order I describe in this book is ultimately agency-based. To address the subject and its agency, or their local manifestations, as an "observable" quantum operator is a possible path towards the solution of the Hard Problem.

Keywords: Aboutness Problem, Agency as the Subject, Decoherence, Field Theories in Consciousness, Fermion, Fractality, Genomic Instantiation, Intelligent Design, Limits of Functionalism, Limits of Computational Theories, Microtubules, Metaphysical, Objective Reduction, Physics, Physical, Problem of Records, Quantum Physics, Subject as Quantum Operator, Subject as Emptiness.

WHY IS A QUESTION ABOUT RECORDS IMPORTANT?

Why does chocolate evoke a subjective flavor of "chocolate" instead of "coconut"?

Paulo J. Negro

Rich combinations of odorants will stimulate olfactory receptors and, in combination with the stimulation of taste bud receptors, become a signal that will be experienced as either "chocolate" or "coconut". Nonetheless, how does the matching between a signal "chocolate" and an experience "chocolate" happen? Where is it written that the signal will match the experience "chocolate", rather than "coconut"? In other words, where is the location of a matching registry for qualia?

As a thought experiment, let us consider that God, or a supreme informational principle, created a universal qualia spreadsheet containing all possible matches between human neuronal activity and human subjective experiences. Where would she have archived this spreadsheet? The question raised here does not contemplate why human brains generate subjective experiences, but what rules match subjective experiences to objective brain states. Admittedly, the specific quality of a quale is illusory, noise. If subjective experiences are randomly assigned to specific patterns of brain activity, where is the randomizer archived? If, on the other hand, matching rules do exist, how can the specific quality of a quale be noise?

Quantum theory, from an information theoretic perspective provides an attractive approach for the problem of consciousness. Record keeping remains a difficult problem in quantum mechanics. A system decoheres as a result of coupling with the environment. Physically, decoherence indicates that records exist[355]. Decoherence can be seen as the loss of information from a system into the environment. Which property of the environment carries the information about the decohered system? How is the amount of information stored by the environment related to the nature or degree of system decoherence [270]?

I objected to the anthropomorphizing of quantum information implicitly proposed by the Orch OR theory. Instead, the state wave collapse can be seen as a form of decodification that depends on the referential context of living systems[356].

For the sake of argument [43], let us consider that coherent quantum oscillations within microtubules and the consequent emergence of quantum states in the cytoskeleton play a role in ionic signal transduction. As a result, deeper rhythms underlying subneural information processing may reflect bidirectional signaling between membrane-bound receptors, ion channels, and the neuronal cytoskeletal matrix [271]. Cortical feedback and feedforward dynamics across spatiotemporal scales will find their projections in putative noncomputational intracellular quantum phenomena.

Hameroff and Penrose [43] point towards a modicum of consciousness for each quantum reduction. However, one should ask: what fundamental property turns living things into observers that are separate and distinct from the environment for decoherence? What property makes the state wave collapse respond to living things with subjectivity, and not with physical information such as position and momentum? Current neuroscience suggests that top-down, not bottom-up processes drive conscious experiences. Thus, the departure point of a quantum theory of consciousness is not the objective reduction of an axiomatic proto-consciousness, but the top-down actions that create a decoherence environment that favors the self-referential transfer of information. An abstract observer emerges from relational processes. Equally abstract, the decodification cannot be measured directly.

Models of consciousness frequently imply the existence of physical fields [272, 273] that convey awareness, including, for example, the search for the origin of consciousness in the interaction among subcellular energy fields. The top-down paradigm applies when photons are considered the fundamental entities that allow for the storage of information[357] which Sorli and colleagues associated to microtubules [274] and Cacha and Poznanski associated to a biophoton field, realized as a conscious field at the level of DNA organization[358] [275]. Cacha and Poznanski [275] see subjective experiences as resonant phenomena that emerge as a functional field from the matching of top-down expectations and bottom-up activity.

Any reciprocal associative network receiving signals from feedback loops may enact a looping time expressed as phase synchrony[359]. The looping time of signals propagating from the dendritic tree to the axonal transport, may allow for the mapping of the whole onto each of its parts [276] as a form of fractal image compression.

Bieberich [277] sees calcium waves traveling from the dendritic tree to the soma as giving origin to a geometrical fractal arrangement of biological material. He further hypothesizes this as the origin of a multidimensional particle capable of making matter conscious. The fractal compression is given by an iterative functional system that repeats the contraction mapping operation to progressively smaller sizes[360]. Putative fractal compression in recursive networks is amenable to the strange loop *expectation.information* treatment reported in this book.

Albeit attractive, quantum theory has limitations. Quantum theory does not solve the 'aboutness' problem [55], much like computational theories. Moreover, conscious experiences emerge from spatiotemporal patterns of integrated activity of different brain areas at a level that quantum theory cannot address:

> *"...the temporally joint action of several tens to hundreds of neurons (organized spatially in a local assembly) is minimally necessary to understand and explain the emergent cognitive and conscious phenomena"* [54].

A similar pattern holds true for events taking place at the mesoscopic level of brain organization, characterized by *"computations and information transfer in the cortex accomplished via* spatial-temporal patterns of functionally related neurons." Such connectivity can be described topographically. Oscillations of neuronal assemblies tuned to the same frequency ranges unfold in spatiotemporal patterns:

> *"Neurons which are tuned to respond to the same feature of a complex stimulus are engaged in an episode of synchronous activity and thereby identify their activity as part of a population-coded signal produced by the distributed processing of a unified assembly"* [54].

In other words, the emergent properties that allow for the conscious representation of objects, and the development of invariant conscious "knowing" require stable resonance of neuronal activity much beyond the levels of organization addressed by quantum physics. Conscious experiences require multiple computational reprocessing [162]. The Orch OR proposes a transition of quantum phenomena into physical events by engaging microtubules and neuronal activity. The theory does not address how neuronal activity subsequently translates into the phenomenology of conscious objects, such as odor-objects, at higher functional levels of brain activity.

THE PROBLEM OF RECORDS IS PHYSICAL, NOT METAPHYSICAL

The question about records addresses the problem of matching neural signals with specific quale, which requires an explanation based on the physics of our universe. It is a specific non-metaphysical aspect of the Hard Problem.

Metaphysical aspects of conscious experiences have a relational abstract origin. Experiences such as love and friendship, or concepts such as ethics and ideals are not measurable in the natural world. Neither are qualia. The "greenest" of green is not measurable. Nonetheless, the matching of "greenest" to green requires a formal physical explanation. The problem of records is relevant because it does not relate to a question about "why", but to a question about "where".

THE PROBLEM OF RECORDS IS NOT ALGORITHMICALLY COMPRESSIBLE

I parsed the Hard Problem of consciousness in two components: an operation for the realization of awareness and a mechanism for the generation of the content of awareness. Put differently, consider the computational origin for mechanisms of awareness:

> *"The existence of regularities may be expressed by saying that the world is algorithmically compressible ...Viewed this way, the question 'Why is the universe knowable?' reduces to 'Why is the universe algorithmically compressible?' and 'Why are human beings so adept at discovering the compressions?' First it should be noted that our mental model of the world is itself an algorithmic compression. If the world were not compressible this way, there could be no cognition. So, the very fact that we exist as observers already constrains the universe to have the property of algorithmic compressibility. Of course, this anthropic reasoning does not constitute an explanation of why the universe is compressible; it merely tells us that we could not be around to debate the issue were it not so"* [278].

I previously discussed how perceptual processes represent emergent properties of systems that conform to the free-energy principle. The minimization of surprise in exchanges with the environment under expectations encoded by state or configuration [72] underlies the common origin of the structural order seen in living systems. It also ultimately led to the emergence of human consciousness from a hierarchical Bayesian brain organization. Thus, mechanisms of awareness, including abstract metaphysical awareness, have a computational foundation based on the incorporation of causal relationships in the environment[361].

Record keeping itself poses a difficult problem in quantum mechanics. As a result, the matching process lies at the heart of a noncomputational origin of consciousness. Both abstract awareness and qualia-laden subjectivity share the same origin: information that becomes realized as subjectiveness from a "me-ness" perspective.

The evolution of the first single-celled organisms provided an information platform that would later become consciousness. The boundary separating the cell and the environment is not metaphysical[362]. This blueprint emphasizes that the "general knowledge" of consciousness is relational. It originates from potential states. For this reason, it is something out of nothing. The paradigm *information.expectation* underlies consciousness at the lowest and most general level of organization[363].

Reimann and colleagues used algebraic topology as a framework to describe the emergent behavior of cortical neural networks in terms of their underlying structure [279]. Algebraic topology provides a mathematical tool that quantitatively describes local and global network properties. These authors represented the synaptic network as a directed graph with neurons as its vertices and synaptic connections from pre- to postsynaptic neurons as its edges[364]. They established the direct relationship between the *"structural graph and the emergent flow of information in response to stimuli, as captured through time series of functional graphs."* This well-designed study combined mathematical modeling with empirical support from cortical and thalamic networks. It provides support to the idea of consciousness as an abstract realization of information described in this blueprint:

> *"We verified the existence of high-dimensional directed simplices[365]in actual neocortical tissue. We further found similar structures in a nervous system as phylogenetically different as that of the worm C.elegans, suggesting that the presence of high dimensional topological structures is a general phenomenon across nervous systems".*

Functional organization occurs in spatial locations of the cortical circuitry consistent with those described at the chapter reviewing signals of consciousness at layer 5[366] and was associated with the highest dimensional directed simplices at layers 5 and 6. These findings indicate how the brain realizes information across dimensions and neuronal ensembles, such as described in the chapter on hippocampal place cells and place fields:

> *"We showed that the spike correlation of a pair of neurons strongly increases with the number and dimension of the cliques they belong to and that it even depends on their specific position in a directed clique. ...These observations indicate that the emergence of correlated activity mirrors the topological complexity of the network. ...The fact that each neuron belongs to many directed cliques of various dimensions explains in vivo observation that neurons can flexibly join multiple ensembles"* [279].

The neocortex seems to process information through a mathematically described sequence of *"cavity formation and disintegration in response to stimuli"*, which binds neurons into cliques of *"increasingly higher dimension as a specific class of cell assemblies, possibly to represent features of the stimulus, and by binding these cliques into cavities of increasing complexity, possibly to represent the associations."* This mechanism lends support to the idea of conscious experiences that emerge in high-dimensional abstract active cliques and become realized across functional hierarchies.

COMPETING EXPLANATIONS OF CONSCIOUSNESS IMPLY DIFFERENT VERSIONS OF THE UNIVERSE

Explanations about the origin of consciousness can have two different narratives.

A universe where locally encoded information realized in a global workspace provides a computational cause for consciousness differs fundamentally from a universe where consciousness emerges from the noncomputational realization of information. In the former, mathematics drives reality and abstraction causes consciousness. In the latter, consciousness emerges from information realized in the subcellular matter-energy.

Is consciousness a consequence of mathematics? I take Wheeler's position, that the question of consciousness is better approached as information-theoretical [278], *"a useful assumption for physical things represented by mathematical descriptions and measurements"* that can be transposed to subjectivity:

> *"John Archibald Wheeler used to write physics equations on the blackboard, stand back, and say, 'Now, I'll clap my hands and a universe will spring into existence.' Of course, it didn't. Stephen Hawkins asked in A Brief History of Time, 'What is it that breathes fire into the equations and makes a universe for them to describe?' Such utterances reveal the absurdity of the view that mathematics is prior to nature. Math in reality comes after nature. It has no generative power"* [278].

THE ABSENT SUBJECT

Although strange loops may provide a mechanism for the generation of awareness, they do not directly explain the subject who experiences awareness. Attempts to cross this last mile may prove fruitless for a long time, as the subject is empty, it is what is not there. It has been argued that functionalism is an insufficient account of the mind [280]. I argue here that functionalism is ultimately empty of the subject.

Imagine a field of nothingness or, perhaps more appropriately, a relational order that is not materially present except for the physical parts that produce relationships. It is thrice empty. It is empty because at higher levels the brain generates computational abstractions describing phenomenal objects. It is empty due to its relational origin. It is empty because the *expectation.information* strange loop stems from potential states.

An alternative interpretation [281, 282] for the emptiness of this abstract order emerges from the idea that consciousness could have a physical, but not observable, presence in the matter of extra-dimensions. This approach would have the universe embedded in a higher dimensional bulk space-time and models consciousness by fermion interactions. Nonetheless, physical models still must be reconciled with a relational origin of consciousness and the integration of top-down and bottom-up streams as indicated by current neuroscience.

In this book I conceived a variable Q to describe the information conveyed at this order and compared it to imaginary numbers. This order cannot be measured, as it is not physical. Nonetheless, it connects the global brain computational workspace to subcellular structures that have become systems that, in turn, have become information. As described above, the iterative mapping of information from higher order brain structures to smaller cellular and molecular levels can be described as a form of "fractalization" of information [277].

Can this relational order of emptiness, this void, function as an environment for decoherence?

As previously stated, Orch OR proponents anthropomorphize quantum information. That quantum information should provide a source of consciousness stems from circular thinking. The observer in quantum experiments interferes with the results, thus quantum state reduction should correspond to consciousness. It is a cryptic intelligent design mistake similar to the more obvious one made by IIT proponents. The subject must precede integrated information and objective reductions.

Quantum theory drives the Hard Problem to another dimension, but the question remains: why consciousness arises from quantum states? In decoherence, a particle will carry information about position and momentum. A quantum system conveys information about position and momentum to a particle. But what happens when the environment for decoherence includes a relational order that is abstract, and nonetheless still there?

This relational order is by strange loop design self-modeling because – as per Gödel – it refers to itself. Natural selection constraints shaped such relational order to infer hidden causes behind events in the environment. It is an inferential relational order. Could this environment for decoherence ultimately transform decoherence into information about a subject, and thus information about consciousness?

Ultimately, the problem becomes a question of records – the fundamental property that may turn living things into observers that are separate and distinct from an environment for decoherence. If a mechanism exists for decoding quantum information based on this shadowy relational order, then a connection may be established between quantum theory and brain computation. Of course all of this is quite speculative, but so are quantum hypotheses for consciousness. Nonetheless, these strange loops originate in actions. Neural Correlates of Consciousness identify what the subject does. Predictions in the neurology of Predictive Coding are active expectations.

Active expectations can only exist as the consequence of an organism's agency. Natural selection has no agency, but its processes created organisms with agency. Thus, this relational order is ultimately agency-based. The subject is agency based. Better said, agency can be seen as the subject. I wonder if agency can make the relational order a field for objective reduction rather than decoherence.

> *"Quantum approaches, on the other hand, have come to represent observables not as noun-like variables, passively possessed by equations, but as operators, acting on those functions as symbolic agents of experimental measurement. The quantum operator imbues an operated function with that observable 'quality' lying latent in the observable-verb's specific agency of empirical assay and may thereby pre-limit the expected set of possible quantitative readings"* [283].

At this point, I believe that approaching the subject and its agency, or their local manifestations, as an "observable" quantum operator is a possible path towards the solution of the Hard Problem. But this is a problem for future minds to solve.

NOTES

[355] Decoherence bears an intimate connection with the existence of records that correlate with past alternatives: *"...as we saw in the analysis of the quantum Brownian motion model, the records can be stored in the decohering environment, and by inspecting them at the final time we can recover the past history of the system. What the decoherence of the quantum Brownian particle by a thermal environment means, therefore, is that the history of the Brownian particle may be recovered by examining the thermal environment. Similarly, the decoherence of a gravitational field by a decohering matter field environment means that we can recover the history of the gravitational field by examining the matter field at late times, which is indeed exactly what is done in cosmology. From a practical point of view therefore, the significance of decoherence is that it ensures a correlation between present records and past events"* [270].

[356] Hameroff and Penrose's turn to quantum mechanics is understandable [43], even seductive: *"within the realm of quantum information theory, the functional wave equation of standard quantum theory provides continuum idealizations, while concealing the ultimate source of information from which they derive"* [278].

[357] According to Sorli and colleagues [274] microtubules may receive information *"via biophotons from 3D quantum vacuum, an idea that is supported by recent research suggesting that microtubules use a binary system for storing information..."* *"...consciousness is intrinsically connected to the spin process, which provides the linchpin between mind and brain, and emerges from the self-referential collapses of spin states and the unity of mind is achieved by entanglement of these mind-pixels. The starting epistemological foundation of this theory lies in the fact that spin is basic quantum bit ('qubit') for encoding information and, on the other hand, neural membranes and proteins are saturated with nuclear spin carrying nuclei and form the matrice of brain electrical activities"* [274].

[358] The possibility of a genomic instantiation of consciousness brings forward an

interesting question. The relationship genes → protein expression can be seen as a bottom-up *expectation.information* relationship. From an information perspective, neurons see the interaction of two *expectation.information* streams, the bottom-up.

gene.protein expression

and the top-down derived from the synchrony between

neural oscillations.subneuronal oscillations

The interaction between these streams may convey the realization of a strange loop deep within its cellular origins.

[359] This includes phase locked recurrent loops at the anterior piriform cortex ←→ posterior piriform cortex and the dentate gyrus ←→ CA3 pyramidal neurons.

[360] "The Hutchinson operator (H(S)) copies downscales versions of the image S_n onto itself so that the resulting image S_{n+1} is a union of $f_i(S_n)$: $H(S){:}S_{n+1} = \cup f_i(S_n) = f_1(S_1) \cup f_2(S_1) \cup f_3(S_1) \ldots f_N(S_n)$"

[361] *"The similarity between the form or structure of the brain and statistical models means that perceptual inference and learning lends itself nicely to a hierarchical treatment, which considers the brain as an empirical Bayesian device. The dynamics of neurons or populations are driven to minimize error at all levels of cortical hierarchy and implicitly render themselves posterior or conditional modes (i.e.*, most likely values) of the causes of given sensory inputs …a 'higher' area dealing with more abstract information about the world, the other 'lower', dealing with more concrete data. The higher area attempts to fit its abstractions to the data it receives from lower areas by sending back to them from its deep pyramidal cells a template reconstruction best fitting the lower level. The lower area attempts to reconcile the reconstruction of its view that it receives from higher area with what it knows, sending back from its superficial pyramidal cells the features in its data which are not predicted by the higher area …the causal structure of the world is embodied in the backward connections. [Feedforward] connections simply provide feedback conveying prediction errors to higher levels" [72].

[362] The standard model does not capture a universe where "me-ness" is at the origin of non-measurable qualia content. Nonetheless such universe still requires mathematical models that match qualia with neuronal patterns of activity, and archives for this information.

[363] Can it be tested? Possibly yes, and in a not so distant future. Current brain-machine interface experiments use an array of electrodes implanted in different regions of the brain to record local field potentials and neuronal firing in multiple locations, or to stimulate neurons or group of neurons. At this point, much of the work includes algorithms that translate brain activity into the movement of artificial limbs, computer interfaces, or that stimulate the brain. This line of research argues against a strict and precise location for motor function. Moreover, receptive fields and maps represent a *"dynamic and fluid spatiotemporal distribution of potential probabilistic neuronal population firing patterns"* [207]. The technology may allow for the connection of the human brain to a computer running algorithms that simulate inferential processes. These algorithms would simulate the sustained oscillatory activity encoding the ensemble densities $q(\vartheta, \lambda)$ to minimize free-energy. They would further simulate the minimization of free-energy across hierarchical levels through

'expectation.information'expectation.information

A clever study design may take advantage of this simulation self-modeling. It would likely connect the brain through different input and output portals. Direct brain connection may thus make it possible for a human being to know "how it feels" to be a computer running a self-modeling inference-based software. Modifications of the algorithm may test whether changes in simulation, but not changes in hardware, modify this putative additional awareness. Although this may seem far-fetched, direct stimulation of the brain during neurosurgical procedures was also seen as improbable, until Penfield actually performed it [329]. My prediction? We will not know "how it feels" to be a computer. The information exchanged with the human brain will become awareness much as observed in the rubber hand experiment, or it will be added to the corporal schema such as described in tool handling. It will provide extra aspects to ongoing sensorial experiences. The experiment will generate an augmented brain, perhaps capable of obtaining statistics of the surround world by different manners. However, there will be no independent "how the computer feels" accessible to the subjective human experience.

[364] "*Networks are often analyzed in terms of groups of nodes that are all-to-all connected, known as cliques. The number of neurons in a clique determines its size, or more formally, its dimension. Directed cliques fit nicely into directed graphs. Such cliques contain a single source neuron and a single sink neuron and reflect a specific motif of connectivity, wherein the flow of information through a group of neurons has an unambiguous direction. The manner in which directed cliques bind together can be represented geometrically. When directed cliques bind by sharing neurons, but due to missing connections, fail to form a larger clique; they form cavities (holes, voids) in this geometric representation. High-dimensional cavities form when high-dimensional (large) cliques bind together*" [279].

[365] Simplices: the mathematical building blocks of the microcircuitry.

[366] "*The activity started at depths that correspond to the locations of the thalamocortical input, increasing in layer 4 and at the top of layer 6 as β1 [Betti number serving as topological metrics] consistent with the finding that most directed simplices are in these areas. The transition from increasing β3 to increasing β3 coincided with the spread of the upper activity zone deeper into layer 5 and the top of layer 6, consistent with the presence of the highest dimensional directed simplices in these layers. The bottom activity zone also continued moving deeper, until it eventually subsided. As the top activity zone reached the bottom of layer 5, β3 attained its peak. The zones of activity at the peaks of β1 and β3 are highly complementary: zones active during the peak of β1 were generally inactive at the peak of β3 and vice-versa.* The activity zone then remained in layer 5 until the cavities collapse" [279].

CHAPTER 20

Final Conclusions

"All models are false, but some may be useful" [330]

Abstract: Consciousness cannot be directly measured. It is not physically present. It is an abstraction that emerges from the relationships between processes. Natural selection transformed structures into processes with informational value. Processes have expectations. Subjectivity is the realization of expectations built into the processes that connect a subject and the world around it. Qualia become possible when a subject realizes locally encoded information. Inferences of causal regularities in the Bayesian sampler brain originated from the structural order of living systems. The realization of awareness of location arises from the matching of maps generated by hippocampal activation. These maps give origin to omnidirectional cell assemblies that can represent abstract meaning. Actions generate concrete and abstract maps. State-dependent hidden expectations in local networks also underlie active expectations in time perception, in other subjective experiences, and mechanisms for conscious perception that trace back to the early components of the brain decision network. Interoception realization arises from the hierarchical architecture of the insula consistently with instantiation of successively higher re-representations. Interoception likely arises from a strange loop merging a represented body with a simple self in which the represented body is a building block of the self it represents. Predictive Coding lends support to a Bayesian inference of hidden causes in the environment. It is a generative model that depends on actions. Within its context, perception corresponds to the inverse mapping of the hidden causes of sensations. Ultimately, being the subject is the realization of inference inversely mapped out of hidden causes of global, integrated actions. Olfactory perception abstracts regularities from the environment into odor-objects of adaptive value through re-representation of structural maps into categories that arise from iterative neuronal assembly oscillations. Strong evidence suggests that human consciousness arises from globally distributed brain activity enabling sustained coherent oscillatory states. The existence of reentrant thalamocortical and corticothalamic oscillations supports the idea of realization of abstract information through re-reading of the strange loop *'expectation. information' expectation. information* into *'prediction. predictive-error'prediction. predictive-error* and, ultimately, into the *'top-down. bottom-up'top-down. bottom-up* and the *'feedback. feedforward'feedback. feedforward* of cortical and thalamic loops. The ideas contained here are consistent with the empirical evidence provided by the IIT. However, this blueprint moves away from the IIT cryptic intelligent design and panpsychism. The adoption of strange loops as a core component of this blueprint is consistent with the Predictive Coding paradigm, with embodied abstractions generated at the insula, and with a multiple level self-modeling description through which lower levels provide primitives for higher-level modeling. Strange loops reconcile the local and global

Paulo J. Negro

paradox of consciousness. These ideas are consistent with the resonance principles contained in the ART, whereby resonances supporting qualia arise from the realization of information in the strange loop *'expectation.information'expectation.information* in which *'expectation.information'* ⟷ *expectation.information* at higher levels of functional organization.

Keywords: Active Expectations, Adaptive Resonance Theory, Bayesian Sampler, Computational Models, Default Mode Network, Embodied Abstractions, Effective Information, Focal Awareness, Generative Models, Global Neuronal Workspace Theory, Interoception, Integrated Information Theory, Mechanisms of Consciousness, Noncomputational Subject, Omnidirectional Cells, Phenomenal Consciousness, Predictive Coding, Problem of Records, Quantum Theories of Consciousness, Realization of Information, Self-Modeling, Sustained Oscillations.

A BRIEF SUMMARY

The fundamental insight guiding this book is simple: consciousness cannot be directly measured because it is an abstraction arising from relationships between processes. It can spring out of nothing because its strange loop is based on the potential states of these processes.

Processes can have "expectations". A car engine has "expectations" that gas be poured in the chamber, ignited, pistons move, *etc.* A car engine is not aware of these "expectations". Its human engineers designed the car engine expecting certain processes to occur, or better yet, the engine was designed for these processes to occur. The information representing the designers' expectations is embodied in the physical structure of the car engine.

The biology of cellular processes has immensely more complex "expectations". These expectations are a form of knowledge. "Expectations" refer to the relationship between something physically or functionally happening and some potential. However, living processes were not designed for anything. Instead, they emerged out of nothing of their own kind.

Natural selection, acting on the dialectic between the first single-celled organisms and their environment, enabled the transformation of structures into processes with informational value. These mechanisms are manifested through the free-energy principle connecting maintenance of structural order in living processes with inferring causal regularities in the Bayesian brain or a brain that functions as a Bayesian sampler. The proposed paradigm has fewer inherently anthropic characteristics than those put forth in other theories of consciousness.

In view of the matching of *information.expectation* with the concept of effective information underlying the IIT phi-consciousness, this model should be consistent with the IIT empirical evidence. However, it avoids the IIT fundamental flaw and its consequent cryptic intelligent design and panpsychism. This blueprint shares the IIT reliance on effective information because it is based on potential states, but it supersedes the IIT.

How do these ideas fit with the neuroscience review in this book?

As we saw in the section on focal awareness, the realization of awareness of location arises from the matching of maps generated by hippocampal activation. These maps can give origin to maps of omnidirectional cell assemblies that have the ability to represent abstract meaning. Multiple episodes of common junctions can free these neurons from their physical context. As actions generate concrete and abstract maps, the strange loop:

'expectation.information'expectation.information

fits with Alva Noë's[367] and Ezequiel Morsella's[368] view of consciousness as an adaptive action [48, 284]. The strange loop is also consistent with Deacon's relational transformation of information from structures into functional relationships [55, 71]. Finally, it provides support to the idea of hippocampal realization of abstract information of omnidirectional cells:

omnidirectional 'ana.afc'ana.afc[369]

The abstraction achieved by functional decoupling of spatial signal contents and realized as "aboutness" is consistent with the concept of thermodynamic decoupling described in this blueprint.

As I reviewed the realization of awareness of time, I attempted to show the importance of state-dependent hidden expectations in local networks. State-dependent local learning and oscillatory phase shifts support the idea of active expectations in time perception and other conscious experiences as well as evidence of mechanisms for conscious perception tracing down to the early components of the brain decision network.

The review of interoception realization found evidence in support of a hierarchical architecture of the insula consistent with the instantiation of successively higher re-representations. Interoception illustrates the emergence of a basic form of awareness out of strange loops merging a represented body with a simple self in which the represented body is a building block of the self it represents. The insula instantiation of awarenesses is consistent with consciousness as an abstraction emerging from multiple strange loops of hierarchical processual realization.

Predictive Coding is consistent with a Bayesian inference of hidden causes in the environment. It is a generative model dependent on actions in which perception corresponds to the inverse mapping of sensation's hidden causes. Generative models do not carry an intelligent design bias. They are subject-centric. Predictive coding and its generative implications fit the framework of a hierarchy of abstractions underlying the active inference built into the interoception process. It is a top-down model based on actions or intentions, supporting the transition of:

'expectation.information'expectation.information

into

'feedback.feedforward'feedback.feedforward

and

'top-down.bottom-up'top-down.bottom.up

The review of realization of olfactory awareness stressed the crucial transition from external information into organism-centric encoding through the transformation of odorants' structural patterns into odor-objects of adaptive value. Not surprisingly, the piriform cortex shares properties with the hippocampus, including the generation of neuronal ensembles through oscillations, time-dependent plasticity of iterative recurrent networks, and the realization of information achieved by the interaction between wide and narrow patterns of activity respectively delivered by mitral and tufted cells, which is consistent with the principle *expectation.information*.

These findings suggest that olfactory perception abstracts regularities from the environment and conveys information of adaptive value. This process is consistent with qualia that actually provide adaptive information about reality and the use of a variable "Q" for its representation as explained earlier in this blueprint. The olfactory system re-representation - from structure-based categories depicted in spatiotemporal maps of odorant molecular features into odor-object categories that match conscious olfactory experiences - parallels the hierarchy of abstractions described in the insula. These categories of relational origin fit with phenomenal perception and require sustained expectations from recurrent neuronal ensembles.

The section on the globalization of consciousness lists strong experimental evidence in support of a human consciousness that emerges from globally distributed brain activity. The evidence points towards a global neuronal workspace that enables coherent self-sustained states. Network simulations indicate massive non-conscious processing interspaced with serial conscious stepwise decisions.

A corticothalamic "parallel-interactive" organization characterized by small-world architectures forms the basis for the interaction of multiple activity arrays – and for the overlap of multiple spatiotopic maps. The realization of abstract information through the cooperation and competition of these maps is not conceptually different from the realization of information across omnidirectional hippocampal ensembles described in the section of focal awareness.

Reentrant thalamocortical and corticothalamic oscillations process information that allows for conscious access, lending support to the idea of realization of awareness with oscillations as a mechanism for sustained expectations in self-modeling:

'prediction.predictive-error'prediction.predictive-error

which in this context represents the strange loop in the process of realizing information into consciousness.

The GNW theory provides evidence in favor of non-qualia laden abstract realization of information, such as with Feelings of Knowing (FOK). Such evidence is also consistent with the division proposed here for the Hard Problem as a mechanism for realization of awareness, and for the content of awareness, itself.

Studies with general anesthetics point to the importance of the Default Mode Network in states of endogenous consciousness. The thalamus does not play a crucial rule in the Default Mode Network. It may be possible to reconcile a thalamic role in consciousness, as proposed by the GNW theory, and the central position of the DMN by referring to phenomenological descriptions in classic Zen texts of a subtle field of consciousness that is not object-bound. This empty field may be enhanced by non-specific thalamic projections to the apical pyramidal dendrites.

The study of anesthesia suggests a causal role for feedback → feedforward (top-down → bottom-up) in human consciousness. General anesthetics with different pharmacodynamics share the property of disrupting top-down ←→ bottom-up signaling on the apical dendrites of pyramidal neurons ←→ cellular bodies. Waves of calcium-mediated activity seem to convey these signals.

Computational models of attention are particularly relevant because attentive matching of top-down expectations and bottom-up feedforward input lies at the center of the ART. While the GNW theory approaches consciousness dynamics from a horizontal perspective, the ART does the same from a vertical perspective. The theory stresses the matching of top-down expectations with bottom-up information at several reprocessing levels and across several complementary computing processes as the center of those brain computations that generate complete and stable representations of the world. These computations should reduce uncertainty and are thus consistent with the reduction of free-energy or surprise across a hierarchy of abstractions, as addressed in the chapter on statistical physics.

The ART can be considered consistent with the cortical signals of consciousness identified in cortical layer 5. As reviewed, the high-frequency calcium-mediated spike bursts of pyramidal cells signal feedback and feedforward interactions, which promote the association of information from multiple cortical areas. These processes are consistent with the unpacking of:

'expectation.information'expectation.information

as a multiple level self-modeling description in which lower levels provide primitives for higher-level modeling. It is consistent with the Predictive Coding paradigm, with the embodied hierarchy of subjective abstractions in the insula, and with the overall computational model of brain function according to which abstractions utilize lower levels as primitives.

The ART concept of neuronal shroud-activity supporting qualia features is consistent with the realization of information across abstract invariant categories. It provides a feasible approach to reconcile the local and global interaction paradox in consciousness. Information is best seen as distributed in the brain, with no absolute or fixed spatial borders, and as a neuronal space-time continuum under functional constraints.

These ideas are consistent with the resonance principles of the ART, in which qualia-supporting resonances arise from realization of information in the strange loop:

'expectation.information'expectation.information

'in which

'expectation.information' ⟷ *expectation.information*

at higher levels of functional organization, which can also be read as

'prediction.predictive-error' ⟷ *prediction.predictive.error*

in the strange loop

'prediction.predictive-error'prediction.predictive-error

'or

'.feedback.feedforward'feedback.feedforward
'top-down.bottom-up'top-down.bottom-up

resonating at the perceptual object level when

'feedback.feedforward' ⟷ *feedback.feedforward*
'top-down.bottom-up' ⟷ *top-down.bottom-up*

in corticocortical and corticothalamic loops.

The realization of this strange loop as conscious experiences encompasses multiple levels of processing because *expectation.information* of lower levels is taken as a *primitive* and manipulated as "information" at higher levels.

It is intuitive to think that the prediction and predictive-error matching signals perceptual representations of the world capable of supporting the organism integrated, goal directed adaptive behaviors.

A given computational perspective does not necessarily exclude the others. A model that acknowledges a role for information integration, self-modeling, hierarchical representations, a global workspace, and with a solid tie to dynamic oscillatory states will capture the nature of consciousness more comprehensively than individual models.

Nonetheless, these models share the principle of conscious experiences as abstractions realized out of map overlays and provided by sustained oscillatory activity. Applying the generative model of predictive coding to the global interactions of consciousness supports my prediction that "being the subject" is the realization of inference inversely mapped out of hidden causes of global integrated actions.

Finally, considering the clear and convincing evidence that globalization of information is necessary for human consciousness, this blueprint explains a contradictory finding: how anencephalic children may experience a degree of awareness without a functional cortex. Awareness and, ultimately, full human consciousness depend on self-referential strange loops that, at a primordial level, may already be instantiated at the upper brain stem superposition of spatial senses in a pre-motor framework, allowing multimodal integration for action. The superior colliculus supports the generative self-simulation gamma range oscillations that may underlie Predictive Coding and primitive phenomenal experiences.

CONCLUSIONS

Cellular processes have innumerable loops. These loops include transformation of information across different states, such as information in the genetic code, information in the tridimensional format of proteins, information in complex and relational processes.

As portrayed in this blueprint, consciousness is an embodied abstraction that predicts two possible versions of our universe:

a. A universe where high dimensional mathematical objects that realize information can directly generate consciousness;
b. A universe where the underlying organic matter and information transformed into an irreducible single entity in living organisms can directly generate consciousness.

I addressed some aspects of this dichotomy by emphasizing the discussion between computational and noncomputational aspects of consciousness. The levels of functional brain organization necessary for acquiring stable representations of conscious-objects clearly go much beyond the descriptions of quantum phenomena. The emergent behavior of cortical neural networks, described with algebraic topology, displays a flow of information characterized by high dimensional topological structures, which was described as reaching their highest levels at layers 5 and 6.

This pattern is consistent with the idea of realization of information across higher dimensions and across neuronal ensembles, as described in this blueprint. A strange loop based on potential states supports the idea of a consciousness that emerges in high-dimensional abstract cliques and becomes realized across the functional hierarchy of the brain.

However, this is not proof that consciousness spring out of mathematical descriptions. An information-theoretical explanation at the bottom of reality remains attractive. The problem of records may play an important role in the solution of this problem. An algorithm may exist that matches qualia with patterns of neural signals. However, record keeping itself poses a problem that is not algorithmically compressible and therefore not amenable to a computational solution.

Living things consist of complex functional systems. If the observer is abstract, having emerged from relational processes, the carried information will not necessarily be physical, such as position and momentum, and it will not be directly measurable. This order cannot be directly measured, as it is not physical. This relational order is by strange loop design self-modeling because – as per Gödel – it refers to itself.

There are no computations directly describing the subject. Computations describe internally- and externally-bound realized information; what the subject does. The subject emerges noncomputationally. *"There is no object outside the subject, there is no subject outside the object"* [63].

Active expectations can only exist as the consequence of an organism's agency. The subject is action-agency based. Rather than search for consciousness with bottom-up quantum theories, top-down downscaled expectations may provide an alternative explanation for an environment where objective reduction is a consequence, not an axiomatic premise. Equally abstract, the decodification cannot be measured directly. The subject and its agency, or their local manifestations, can conceivably become a quantum operator.

Oscillatory activity can be implicitly self-modeling. It may connect attention and first order representationism in the multimodal sensory integration for action at the superior colliculus interface of phenomenal experience with the higher-order representationism of information encoded at the GNW. It may provide a frame for cross-coupling oscillations that convey both information and expectations across hierarchical levels. Phenomenal experiences conceivably emerge with the onset of sustained oscillatory rhythms depicting sustained expectations in an integrative action-based model.

There is no cryptic intelligent design in the *expectation.information* paradigm. It is a subject-centric paradigm. Because this model takes into account the a priori existence of a subject, it provides a venue for solving the local information – global sharing paradox of consciousness. The adoption of a variable "Q" to depict processes supporting conscious experiences at subphenomenal levels allows the freedom to examine mechanisms of focal awareness and their matching with the globalization of information of conscious states.

Subjectivity is not physically present. It can be explained as the realization of the expectations of processes that connect a subject and the world around it. Realized information depicts the organism in the world: being the subject can be explained as the realization of inference inversely mapped out as hidden causes of globally integrated actions. Therefore, an action-based consciousness can be traced to the emergence of agency[370] in single-celled organisms, which first engineered the transformation of the physical into the relational.

Local networks are empty of qualia. Qualia becomes possible when an agent actively realizes locally encoded information. At the same time, the merging of locally depicted information with globally distributed information is a building block of the subject itself. Consciousness in living creatures becomes possible because neural oscillations convey sustained inference across abstraction levels Default mode network activation and attentional network deactivation, as well as default mode network deactivation and attentional network activation are both seen as manifestations of the subject's experiences. Both support the subject's emergence.

In summary, I proposed consciousness as the realization of expectations of processes that functionally bind an organism to its environment. In the course of preparing this work, I arrived at a computational model for the Hard Problem of consciousness. Despite my former beliefs, a noncomputational explanation does not capture all the aspects of the Hard Problem. Consciousness is an abstraction realized out of the human brain's computational hierarchies. Nonetheless, brain computations have noncomputational origins. I described how consciousness could spring out of nothing of its kind. I reported how to connect the realization of information in single-celled organisms and information realized as human consciousness.

Human brain structure and processes that generate human consciousness evolved from single-celled organisms within a conservative natural history. If cellular processes bear some form of "Q consciousness-information", a conservative natural history predicts processually organized mental states and consciousness within similar nonlinear broadcast of constraints. To act is to be.

SOME PERSONAL NOTES

Master Gettan asked a monk: "Keichū made a hundred carts. If he took off both wheels and removed the axle, what would he make clear about the cart?" [63]

The subject is void, and the mind its avatar.

At a younger age, I was influenced by Benda's [285] "The treason of the intellectuals". His ideas about the responsibilities of thinking classes in moderating the appetites and arrogance of national passions are as current now as they were in 1928. The irony of writing this book never failed me: physicians are not intellectuals. We deal with practical things.

This is not a textbook of consciousness. The neuroscience review targets the fitness of the model and is not intended as a comprehensive description of the circuitry of consciousness. For this reason, the references are capricious.

This book was not born out of an attempt to convince others of some preconceived ideas, but from the efforts to convince myself of the validity (or not) of some of my intuitions about the origins of consciousness. This quite personal origin explains a gauche character and the combination of popular science literature with academic peer reviewed papers. I simply stopped whenever satisfied with the information, whatever its origin. Simple heuristics may be less than optimal to fit previously known data but performs better in predictions [121]. To this end I should give thanks to Dr. Ginger Campbell and her Brain SciencePodcast http://brainsciencepodcast.com/ for rekindling my curiosity.

Frankly, despite the length of this book, I find both its philosophical and neuroscience sections elementary. Nonetheless, the neuroscience section is probably a difficult read for philosophers and the philosophical section superficial for the specialist. And *vice-versa* for those with a biological background. They will probably find the later speculative. The book does not include recent methodological breakthroughs in neuroscience. The physics section is likely non-mathematical platitudes for physicist. Nonetheless, the criticism towards the bottom-up model of the Orch OR is solidly anchored in neurobiology. The whole is more informative than its parts. So far, no specialist solved the Hard Problem. I am a generalist. The strength of the book is also its weakness: the bridge between disciplines. Further expansions would likely have provided only incremental benefit to the text.

I initially prepared a short manuscript of about four pages explaining what I thought to be a simple idea: for consciousness to spring out of nothing, it must stem from strange loops based on potential states. I honestly believed it to be an obvious, almost circular argument.

Moreover, while the brain is concretely empty of consciousness, consciousness as an abstraction is something that must be accepted concretely. A blueprint for the Hard Problem must embrace nothingness, or so I thought.

To my surprise, most people did not understand what I was trying to say.

I became unsure about my intuition. I felt compelled to check if I was wrong. Ergo, the traditional review of neuroscience, the descriptions of feedbacks and feedforwards, *etc.* I found that writing supported clear thinking. This was a time-consuming task. I honestly wanted to find myself in clear error, so I could drop it. All this conceptual thinking is a hazard and a burden. Alas, I did not find any compelling evidence that I had been mistaken. Instead, I found this model to be superior to the Integrated Information Theory of consciousness. It can share its mathematical structure, and it certainly avoids the IIT fatal flaw. Nonetheless, it still requires a rigorous mathematical treatment, improvement of its precision and accuracy, and a more careful matching with the neurobiology of consciousness. At best, these are still embryonal ideas.

Finally, I return here to the question raised at its preface. At some point, it so happened that I had dinner with a friend, an NIH scientist, who asked me: "Have you ever considered that people may find this thread relevant to their spiritual questions" (Fig. **15**)?

Indeed, spiritual questions had crossed my mind. I was profoundly influenced by the Gateless Gate [63]. The idea of consciousness as emptiness has always been a thought of mine. What is an abstraction, other than emptiness? How could something be there and not be there at the same time? This is a question akin to the first koan of the Gateless Gate [63]:

A monk asked Joshu in all earnestness:
"Does a dog have a Buddha nature or not?"
Joshu said, "Mu!" (無) [no-thing]

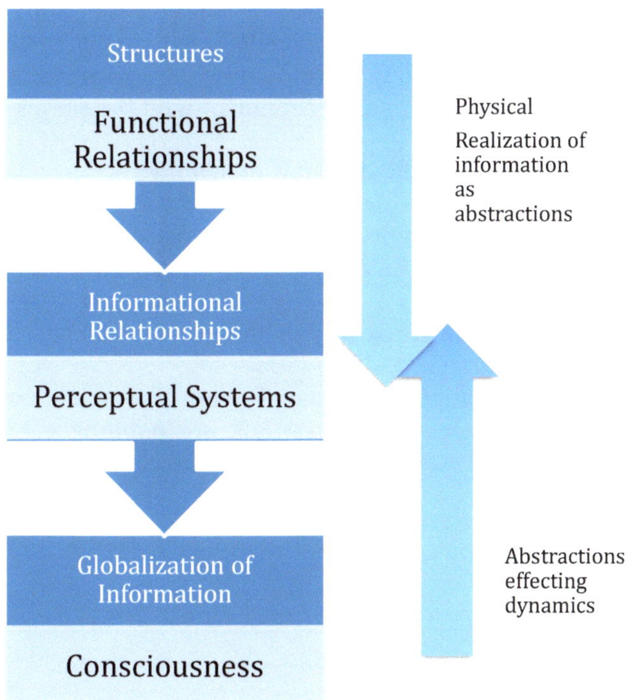

Fig. (15). Turning the thread upside down.

Unexpectedly to me, this work may allow for a new approach to the problem. Something abstract can affect the firing of neurons[371]. I pursued the transition from the physical to the abstract. But the transition can go both ways. The concrete can generate the abstract. Conversely, the abstract can affect the biology of the human brain. Of course, this point brings forward a complicated discussion of causality that escapes the reach of this book.

At the end of the day, the value of ideas only emerges after criticism washes out what is dubious and what is in error.

I believe that it is more likely than not that the ideas in this book point towards the right direction. I hope that the qualities of this book outweigh its limitations:

> *"If you meet a swordsman, you may present a sword;*
> *You should not offer a poem unless you meet a poet.*
> *When you speak to others, say only three-quarters of it;*
> *You should never give the remaining part." [63]*

NOTES

[367] To Noë [48] seeing transcends the content generated by the visual system. "Seeing" is an active adaptive behavior of navigating a tridimensional environment. Blind people reconstruct and navigate the tridimensional environment with the senses of hearing and touch.

[368] To Morsella and colleagues [284], the demands of adaptive action selection require a first-person perspective, which is considered a primitive form of self. Their position implies that the adaptive action in the environment is evidence of a conscious field.

[369] *'abstract- neuronal- assembly. abstract- field- center'abstract- neuronal- assembly. abstract- field- center.*

[370] Agency as a conscious content arises from the correspondence between action intentions and action outcomes.

[371] I am not addressing here the complicated issue of free will. Much of our experience of free will and autonomy of decisions is illusory. I am just pointing out how embodied abstractions capturing phenomenal experiences rule adaptive global actions.

REFERENCES

[1] Shallice T. Dual functions of consciousness. Psychol Rev 1972; 79(5): 383-93. [http://dx.doi.org/10.1037/h0033135] [PMID: 5076868]

[2] Chalmers D. The conscious mind: In search of a fundamental theory. New York: Oxford University Press 1996; p. 3.

[3] Roach J. Journal ranks top 25 unanswered science questions. National Geographic News 2005 June 30; Available from https://www.nationalgeographic.com/

[4] Nagel T. What is it like to be a bat? Philos Rev 1974; 83: 435-50. [http://dx.doi.org/10.2307/2183914]

[5] Block N. On a confusion about a function of consciousness. Behavioral and Brain Sciences 1995; 18: 227-87. [http://dx.doi.org/10.1017/S0140525X00038188]

[6] Treisman AM, Gelade G. A feature-integration theory of attention. Cognit Psychol 1980; 12(1): 97-136. [http://dx.doi.org/10.1016/0010-0285(80)90005-5] [PMID: 7351125]

[7] Schacter DL. Searching for memory: the brain, the mind, and the past. New York: Basic Books 1996.

[8] Squire LR. Memory and brain. New York: Oxford University Press 1987.

[9] Levelt WM. Speaking: From intention to articulation. Cambridge, MA: The MIT Press 1989.

[10] Rosenbaum DA. Motor control. In: Pashler H, Yantis S, Eds. Stevens' handbook of experimental psychology. 3rd. New York: Wiley 2002; 1: pp. 315-39. Sensation and perception

[11] Taylor JA, Ivry RB. Implicit and explicit processes in motor learning.Action science. Cambridge, MA: The MIT Press 2013; pp. 63-87. [http://dx.doi.org/10.7551/mitpress/9780262018555.003.0003]

[12] Lieberman MD. The X- and C-systems: the neural basis of automatic and controlled social cognition.Fundamentals of social neuroscience New York. Guilford 2007; pp. 290-315.

[13] Lashley KS. The behavioristic interpretation of consciousness. Psychological Bulletin 1923; 30 237-272 (Part 1), 329-353 (Part 2)

[14] Einstein A, Infeld L. The evolution of physics. New York: Touchstone 1967.

[15] Gibson JJ. The ecological approach to visual perception. Boston, MA: Houghton-Mifflin 1979.

[16] Merker B. The efference cascade, consciousness, and its self: naturalizing the first person pivot of action control. Front Psychol 2013; 4: 501. [http://dx.doi.org/10.3389/fpsyg.2013.00501] [PMID: 23950750]

[17] Prinz J. The intermediate level theory of consciousness.The Blackwell companion to consciousness. Malden, MA: Blackwell Publishing 2007; pp. 248-60. [http://dx.doi.org/10.1002/9780470751466.ch20]

[18] Morsella E, Godwin CA, Jantz TK, Krieger SC, Gazzaley A. Homing in on consciousness in the nervous system: An action-based synthesis. Behav Brain Sci 2016; 39: e168. [Target Article]. [http://dx.doi.org/10.1017/S0140525X15000643] [PMID: 26096599]

[19] Graziano MA. Consciousness and the social brain. New York: Oxford University Press 2013.

[20] Simons DJ, Levin DT. Change blindness. Trends Cogn Sci (Regul Ed) 1997; 1(7): 261-7. [http://dx.doi.org/10.1016/S1364-6613(97)01080-2] [PMID: 21223921]

[21] Jaynes J. The origin of consciousness in the breakdown of the bicameral mind. New York: Mariner Books 1976.

[22] Bridgeman, personal communication 2014 April 16; [this reference is complete]

[23] Merker B. From probabilities to percepts: a subcortical as locus of phenomenal experience.Being in time: dynamical models of phenomenal experience 2012; 37-80. [http://dx.doi.org/10.1075/aicr.88.03mer]

[24] Cushing D, Gazzaley A, Morsella E. Externally controlled involuntary cognitions and their relations with other representations in consciousness. Conscious Cogn 2017; 55: 1-10. [http://dx.doi.org/10.1016/j.concog.2017.07.003] [PMID: 28735212]

[25] Tsotsos JK. Behaviorist intelligence and the scaling problem. Artif Intell 1995; 75: 135-60. [http://dx.doi.org/10.1016/0004-3702(94)00019-W]

[26] Tsotsos JK. A computational perspective on visual attention. Cambridge, MA: MIT Press 2011. [http://dx.doi.org/10.7551/mitpress/9780262015417.001.0001]

[27] Greene BR. The elegant universe: superstrings, hidden dimensions, and the quest for the ultimate theory. New York: Norton 1999.

[28] Morsella E. The function of phenomenal states: supramodular interaction theory. Psychol Rev 2005; 112(4): 1000-21. [http://dx.doi.org/10.1037/0033-295X.112.4.1000] [PMID: 16262477]

[29] Morsella E, Godwin CA, Jantz TK, Krieger SC, Gazzaley A. Passive frame theory: A new synthesis. Behav Brain Sci 2016; 39: e199. [http://dx.doi.org/10.1017/S0140525X15002812] [PMID: 28355823]

[30] Singer W. Consciousness and neuronal synchronization.The neurology of consciousness. Academic Press 2011; pp. 43-52.

[31] Master Hans Talhoffer. Personal Manuscript 1459; fol2r Wiktenauer: a HEMA alliance project 1459. Available from http://wiktenauer.com/wiki/Main_Page

[32] Kandel ER. From nerve cells to cognition: the internal representations of space and action.Principles of neural science. 5th ed. New York: McGraw Hill 2013; pp. 370-91.

[33] Chalmers D. The conscious mind: in search of a fundamental theory. New York: Oxford University Press 1996.

[34] Nagel T. What is like to be a bat? Philos Rev 1974; 4: 435-50. [http://dx.doi.org/10.2307/2183914]

[35] Block N. On a confusion about the function of consciousness. Behav Brain Sci 1995; 18: 227-87. [http://dx.doi.org/10.1017/S0140525X00038188]

[36] Crick F, Koch C. Consciousness and neuroscience.Essential sources in the scientific study of consciousness. 1st ed. Cambridge: MIT Press 2003; pp. 35-54.

[37] Metzinger T. Being no one. Cambridge: The MIT Press 2003.

[38] Fodor J, Piatelli-Palmarini M. What Darwin got wrong. 1st ed., New York: Farrar, Strauss and Giroux 2010.

[39] Scientific American [home page on the Internet] 2015; 1-4. Oliver Sacks, Antonio Damasio and Others Debate Christof Koch on the Nature of Consciousness. Available from: http://blogs.scientificamerican.com/mind-guest-blog/exclusive-oliver-sacks-antonio-damasio-and-others-debate-christof-koch-on-the-nature-of-consciousness/

[40] Ballard DH. Brain computation as hierarchical abstraction. Cambridge: MIT Press 2015. [http://dx.doi.org/10.7551/mitpress/9780262028615.001.0001]

[41] Emes RD, Pocklington AJ, Anderson CN, *et al.* Evolutionary expansion and anatomical specialization of synapse proteome complexity. Nat Neurosci 2008; 11(7): 799-806.

[http://dx.doi.org/10.1038/nn.2135] [PMID: 18536710]

[42] Tononi G. Consciousness as integrated information: a provisional manifesto. Biol Bull 2008; 215(3): 216-42.
[http://dx.doi.org/10.2307/25470707] [PMID: 19098144]

[43] Hameroff S, Penrose R. Consciousness in the universe: a review of the 'Orch OR' theory. Phys Life Rev 2014; 11(1): 39-78.
[http://dx.doi.org/10.1016/j.plrev.2013.08.002] [PMID: 24070914]

[44] Cartwright J. Evolution and human behavior.Cambridge: The MIT Press 2008; pp. 115-43.

[45] Mandler G. Consciousness: respectable, useful, and probably necessary.Essential sources in the scientific study of consciousness. Cambridge: The MIT Press 2003; pp. 15-34.

[46] Dehaene S. Consciousness and the brain: deciphering how the brain codes our thoughts. New York: Penguin Books 2014.

[47] Zylberberg A, Fernández Slezak D, Roelfsema PR, Dehaene S, Sigman M. The brain's router: a cortical network model of serial processing in the primate brain. PLOS Comput Biol 2010; 6(4): e1000765.http://journals.plos.org/ploscompbiol/article?id=10.1371/journal.pcbi.1000765
[http://dx.doi.org/10.1371/journal.pcbi.1000765] [PMID: 20442869]

[48] Nöe A. Out of our heads: why you are not your brain, and the other lessons from the biology of consciousness. New York: Hill and Wang 2009.

[49] Gödel K. On formally undecidable propositions of Principia Mathematica and related systems. New York: Dover Publications 1992.

[50] Nagel E, Newman JR. Gödel's Proof. New York: New York University Press 2001.

[51] Penrose R. The emperor's new mind. Oxford: Oxford University Press 1999.

[52] Franzen T. Gödel's Theorem: an incomplete guide to its use and abuse. Wellesley: CRC Press 2005.
[http://dx.doi.org/10.1201/b10700]

[53] Cicurel R, Nicolelis ML. The relativistic brain: how it works and why it cannot be simulated by a Turing machine. Natal: Kios Press 2015.

[54] Fingelkurts AA, Fingelkurts AA, Neves CFH. Natural world physical, brain operational, and mind phenomenal space-time. Phys Life Rev 2010; 7(2): 195-249.
[http://dx.doi.org/10.1016/j.plrev.2010.04.001] [PMID: 20417160]

[55] Deacon T. Incomplete nature: how mind emerged from matter. New York: WW Norton 2013.

[56] Posner MI, Warren RE. Traces, concepts, and conscious constructions.Coding processes in human memory. Washington, DC: Winston 1972.

[57] Cerullo MA. The Problem with Phi: A Critique of Integrated Information Theory. PLOS Comput Biol 2015; 11(9): e1004286.http://journals.plos.org/ploscompbiol/article?id=10.1371/ journal.pcbi.1004286
[http://dx.doi.org/10.1371/journal.pcbi.1004286] [PMID: 26378789]

[58] Oizumi M, Albantakis L, Tononi G. From the phenomenology to the mechanisms of consciousness: Integrated Information Theory 3.0. PLOS Comput Biol 2014; 10(5): e1003588.http:// journals.plos.org/ploscompbiol/article?id=10.1371/journal.pcbi.1003588
[http://dx.doi.org/10.1371/journal.pcbi.1003588] [PMID: 24811198]

[59] De Mol L. Generating, solving and the mathematics of Homo Sapiens. E. Post's views on computation.A computable universe: understanding and exploring nature as computation. Singapore: World Scientific 2013; pp. 23-44.

[60] Sieg W. Axioms for computability: do they allow a proof of Church's thesis? In: Zenil H, Ed. A computable universe: understanding and exploring nature as computation. Singapore: World Scientific 2013; pp. 99-123.

[61] Reggia JA. The rise of machine consciousness: studying consciousness with computational models. Neural Netw 2013; 44: 112-31. [http://dx.doi.org/10.1016/j.neunet.2013.03.011] [PMID: 23597599]

[62] Hameroff S, Penrose R. Orchestrated reduction of quantum coherence in brain microtubules: a model for consciousness. Math Comput Simul 1996; 40: 453-80. [http://dx.doi.org/10.1016/0378-4754(96)80476-9]

[63] Yamada K. The Gateless Gate: the classic book of zen koans Sommerville. Wisdom Publications 2004.

[64] Hofstadter D. I am a strange loop. New York: Basic Books 2007.

[65] Futuyma DJ. Evolution. 3rd ed., Sunderland: Sinauer Associates 2013.Sunderland: Sinauer Associates 2013.

[66] Nikolić D. Practopoiesis: or how life fosters a mind. J Theor Biol 2015; 373: 40-61. [http://dx.doi.org/10.1016/j.jtbi.2015.03.003] [PMID: 25791287]

[67] Changeux J-P. The physiology of truth: neuroscience and human knowledge. Cambridge: The Belknap Press 2002.

[68] Changeux J-P. Climbing brain levels of organization from genes to consciousness. Trends Cogn Sci (Regul Ed) 2017; 21(3): 168-81. [http://dx.doi.org/10.1016/j.tics.2017.01.004] [PMID: 28161289]

[69] De la Fuente IM, Cortes JM, Pelta DA, Veguillas J. Attractor metabolic networks. PLoS One 2013; 8(3): e58284.http://journals.plos.org/plosone/article?id=10.1371/journal.pone.0058284 [http://dx.doi.org/10.1371/journal.pone.0058284] [PMID: 23554883]

[70] Pierce JR. An introduction to information theory: symbols, signals and noise. 2nd ed., New York: Dover Publications 1980.

[71] Deacon TW. What is missing from theories of information.Information and the nature of reality: from physics to metaphysics. Cambridge: Cambridge University Press 2010; pp. 146-69. [http://dx.doi.org/10.1017/CBO9780511778759.008]

[72] Friston KJ, Stephan KE. Free-energy and the brain. Synthese 2007; 159(3): 417-58.https://www.ncbi.nlm.nih.gov/pmc/articles/PMC2660582/ [http://dx.doi.org/10.1007/s11229-007-9237-y] [PMID: 19325932]

[73] Madsen PL, Hasselbalch SG, Hagemann LP, *et al.* Persistent resetting of the cerebral oxygen/glucose uptake ratio by brain activation: evidence obtained with the Kety-Schmidt technique. J Cereb Blood Flow Metab 1995; 15(3): 485-91. [http://dx.doi.org/10.1038/jcbfm.1995.60] [PMID: 7714007]

[74] Kuzawa CW, Chugani HT, Grossman LI, *et al.* Metabolic costs and evolutionary implications of human brain development. Proc Natl Acad Sci USA 2014; 111(36): 13010-5.http://www.pnas.org/content/111/36/13010.long [http://dx.doi.org/10.1073/pnas.1323099111] [PMID: 25157149]

[75] Solomon EP, Martin C, Martin D, Berg LR, Eds. Biology. 10th ed., Stanford: Cengage Learning 2014.

[76] Cravatt B, Mackie K, Piomelli D. Understanding the endocannabinoid system. Proceedings of the National Institutes of Health Marijuana and Cannabinoids Neuroscience Research Summit. Bethesda, Maryland. 2016.

[77] Balduzzi D, Tononi G. Qualia: the geometry of integrated information. PLOS Comput Biol 2009; 5(8): e1000462. [http://dx.doi.org/10.1371/journal.pcbi.1000462] [PMID: 19680424]

[78] Buzsáki G. Rhythms of the brain. New York: Oxford University Press 2006. [http://dx.doi.org/10.1093/acprof:oso/9780195301069.001.0001]

[79] Samsonovich A, McNaughton BL. Path integration and cognitive mapping in a continuous attractor neural network model. J Neurosci 1997; 17(15): 5900-20. [http://dx.doi.org/10.1523/JNEUROSCI.17-15-05900.1997] [PMID: 9221787]

[80] Bos JJ, Vinck M, van Mourik-Donga LA, Jackson JC, Witter MP, Pennartz CMA. Perirhinal firing patterns are sustained across large spatial segments of the task environment. Nat Commun 2017; 8: 15602.https://www.nature.com/articles/ncomms15602 [http://dx.doi.org/10.1038/ncomms15602] [PMID: 28548084]

[81] Stensola T, Moser EI. Grid cells and spatial maps in entorhinal cortex and hippocampus.Micro-, meso- and macro-dynamics of the brain. Heidelberg: Springer 2016; pp. 59-80. [http://dx.doi.org/10.1007/978-3-319-28802-4_5]

[82] Buzsáki G, Wang X-J. Mechanisms of gamma oscillations. Annu Rev Neurosci 2012; 35: 203-25. [http://dx.doi.org/10.1146/annurev-neuro-062111-150444] [PMID: 22443509]

[83] Postle BR. The hippocampus, memory, and consciousness.The neurology of consciousness: cognitive science and neuropathology. 2nd ed. San Diego: Academic Press 2016; pp. 349-63. [http://dx.doi.org/10.1016/B978-0-12-800948-2.00021-2]

[84] Dresp-Langley B, Durup J. A plastic temporal brain code for conscious state generation. Neural Plast 2009; 2009: 482696.https://www.hindawi.com/journals/np/2009/482696/ [http://dx.doi.org/10.1155/2009/482696] [PMID: 19644552]

[85] Merchant H, de Lafuente V. Introduction to the neurobiology of interval timing.Neurobiology of interval timing. New York: Springer 2014; pp. 1-16. [http://dx.doi.org/10.1007/978-1-4939-1782-2_1]

[86] Gupta DS, Merchant H. Editorial: understanding the role of the time dimension in the brain information processing. Front Psychol 2017; 8(article 240): 5-8. Available from https://www.ncbi.nlm.nih.gov/pmc/articles/PMC5322218/ [http://dx.doi.org/10.3389/fpsyg.2017.00240]

[87] Buhusi CV, Meck WH. What makes us tick? Functional and neural mechanisms of interval timing. Nat Rev Neurosci 2005; 6(10): 755-65. [http://dx.doi.org/10.1038/nrn1764] [PMID: 16163383]

[88] van Rijn H, Gu B-M, Mecj WH. Dedicated clock/timing-circuit theories of time perception and timed performance.Neurobiology of interval timing. New York: Springer 2014; pp. 75-99. [http://dx.doi.org/10.1007/978-1-4939-1782-2_5]

[89] Kononowicz TW, van Wassenhove V. In search of oscillatory traces of the internal clock. Front Psychol 2017; 7(article 224): 9-12. Available from https://www.frontiersin.org/articles/10.3389/fpsyg.2016.00224/full

[90] Johnston A, Arnold DH, Nishida S. Spatially localized distortions of event time. Curr Biol 2006; 16(5): 472-9. [http://dx.doi.org/10.1016/j.cub.2006.01.032] [PMID: 16527741]

[91] Merchant H, Harrington DL, Meck WH. Neural basis of the perception and estimation of time. Annu Rev Neurosci 2013; 36: 313-36. [http://dx.doi.org/10.1146/annurev-neuro-062012-170349] [PMID: 23725000]

[92] Buonomano DV. Neural dynamics based timing in the subscond to seconds range.Neurobiology of interval timing. New York: Springer 2014; pp. 101-17.

[93] Kosem A. Cortical oscillations as temporal reference frames for perception. PhD Dissertation Universite Pierre et Marie Curie - Paris VI 2014. Available from: https://tel.archives-ouvertes.fr/te--01069219

[94] Ruhnau P, Hauswald A, Weisz N. Investigating ongoing brain oscillations and their influence on conscious perception - network states and the window to consciousness. Front Psychol 2014; 5(article

1230): 1-9. Available from https://www.frontiersin.org/articles/10.3389/fpsyg.2014.01230/full

[95] Finnerty GT, Shadlen MN, Jazayeri M, Nobre AC, Buonomano DV. Time in Cortical Circuits. J Neurosci 2015; 35(41): 13912-6.http://www.jneurosci.org/content/35/41/13912 [abstract]. [http://dx.doi.org/10.1523/JNEUROSCI.2654-15.2015] [PMID: 26468192]

[96] Stefanics G, Hangya B, Hernádi I, Winkler I, Lakatos P, Ulbert I. Phase entrainment of human delta oscillations can mediate the effects of expectation on reaction speed. J Neurosci 2010; 30(41): 13578-85. [http://dx.doi.org/10.1523/JNEUROSCI.0703-10.2010] [PMID: 20943899]

[97] Lakatos P, Karmos G, Mehta AD, Ulbert I, Schroeder CE. Entrainment of neuronal oscillations as a mechanism of attentional selection. Science 2008; 320(5872): 110-3. [http://dx.doi.org/10.1126/science.1154735] [PMID: 18388295]

[98] Busch NA, Dubois J, VanRullen R. The phase of ongoing EEG oscillations predicts visual perception. J Neurosci 2009; 29(24): 7869-76. [http://dx.doi.org/10.1523/JNEUROSCI.0113-09.2009] [PMID: 19535598]

[99] Engel A, Fries P. Neuronal oscillations, coherence, and consciousness.The neurology of consciousness: cognitive science and neuropathology. 2nd ed. San Diego: Academic Press 2016; pp. 49-60. [http://dx.doi.org/10.1016/B978-0-12-800948-2.00003-0]

[100] Craig AD. How do you feel--now? The anterior insula and human awareness. Nat Rev Neurosci 2009; 10(1): 59-70. [http://dx.doi.org/10.1038/nrn2555] [PMID: 19096369]

[101] Craig AD. Significance of the insula for the evolution of human awareness of feelings from the body. Ann N Y Acad Sci 2011; 1225: 72-82. [http://dx.doi.org/10.1111/j.1749-6632.2011.05990.x] [PMID: 21534994]

[102] Gu X, Hof PR, Friston KJ, Fan J. Anterior insular cortex and emotional awareness. J Comp Neurol 2013; 521(15): 3371-88. [http://dx.doi.org/10.1002/cne.23368] [PMID: 23749500]

[103] Simmons WK, Avery JA, Barcalow JC, Bodurka J, Drevets WC, Bellgowan P. Keeping the body in mind: insula functional organization and functional connectivity integrate interoceptive, exteroceptive, and emotional awareness. Hum Brain Mapp 2013; 34(11): 2944-58. [http://dx.doi.org/10.1002/hbm.22113] [PMID: 22696421]

[104] Critchley HD, Garfinkel SN. Interactions between visceral afferent signaling and stimulus processing. Front Neurosci 2005; 9(article 286): 1-9. Available from https://www.ncbi.nlm.nih.gov/pmc/articles/PMC4550795/ [http://dx.doi.org/10.3389/fnins.2015.00286]

[105] Damasio A. The feeling of what happens: body and emotion in the making of consciousness. New York: Harcourt Brace 1999.

[106] Yan L, Wang DJ, Lapidus RC, *et al.* How the heart speaks to the brain: neural activity during cardiorespiratory interoceptive stimulation. Proceedings of the Biological Psychiatry 71st Annual Scientific Convention and Meeting. 4S.

[107] Damasio A. Self comes to mind. New York: Pantheon Books 2010.

[108] Seth AK. Interoceptive inference, emotion, and the embodied self. Trends Cogn Sci (Regul Ed) 2013; 17(11): 565-73. [http://dx.doi.org/10.1016/j.tics.2013.09.007] [PMID: 24126130]

[109] Seth A. How your brain hallucinates reality - [cited 25th Feb 2018] 2018. Available from https://www.ted.com/talks/anil_seth_how_your_brain_hallucinates_your_conscious_reality

[110] Adams RA, Shipp S, Friston KJ. Predictions not commands: active inference in the motor system.

Brain Struct Funct 2013; 218(3): 611-43.
[http://dx.doi.org/10.1007/s00429-012-0475-5] [PMID: 23129312]

[111] Barrett LF, Simmons WK. Interoceptive predictions in the brain. Nat Rev Neurosci 2015; 16(7): 419-29.
[http://dx.doi.org/10.1038/nrn3950] [PMID: 26016744]

[112] Bastos AM, Usrey WM, Adams RA, Mangun GR, Fries P, Friston KJ. Canonical microcircuits for predictive coding. Neuron 2012; 76(4): 695-711.
[http://dx.doi.org/10.1016/j.neuron.2012.10.038] [PMID: 23177956]

[113] Barbas H. General cortical and special prefrontal connections: principles from structure to function. Annu Rev Neurosci 2015; 38: 269-89.
[http://dx.doi.org/10.1146/annurev-neuro-071714-033936] [PMID: 25897871]

[114] Barbas H, Rempel-Clower N. Cortical structure predicts the pattern of corticocortical connections. Cereb Cortex 1997; 7(7): 635-46.
[http://dx.doi.org/10.1093/cercor/7.7.635] [PMID: 9373019]

[115] Chanes L, Barrett LF. Redefining the role of limbic areas in cortical processing. Trends Cogn Sci (Regul Ed) 2016; 20(2): 96-106.
[http://dx.doi.org/10.1016/j.tics.2015.11.005] [PMID: 26704857]

[116] Suzuki K, Garfinkel SN, Critchley HD, Seth AK. Multisensory integration across exteroceptive and interoceptive domains modulates self-experience in the rubber-hand illusion. Neuropsychologia 2013; 51(13): 2909-17.
[http://dx.doi.org/10.1016/j.neuropsychologia.2013.08.014] [PMID: 23993906]

[117] Barrett LF. The theory of constructed emotion: an active inference account of interoception and categorization. Soc Cogn Affect Neurosci 2017; 12(1): 1-23.
[PMID: 27798257]

[118] Kahneman D. Thinking, fast and slow. New York: Farrar, Straus and Giroux 2011.

[119] Gigerenzer G, Hertwig R, Pachur T. Introduction.Heuristics: the foundations of adaptive behavior. New York: Oxford University Press 2011; pp. xvii-xxiii.
[http://dx.doi.org/10.1093/acprof:oso/9780199744282.001.0001]

[120] Sanborn AN, Chater N. Bayesian brains without probabilities. Trends Cogn Sci (Regul Ed) 2016; 20(12): 883-93.
[http://dx.doi.org/10.1016/j.tics.2016.10.003] [PMID: 28327290]

[121] Gigerenzer DG, Brighton H. Homo heuristicus: why biased minds make better inferences.Heuristics: the foundations of adaptive behavior. New York: Oxford University Press 2011; pp. 2-30.
[http://dx.doi.org/10.1093/acprof:oso/9780199744282.003.0001]

[122] Gould SJ. Not necessarily a wing [cited: 23rd February 2018] 2018. Available from http://www.stephenjaygould.org/library/gould_functionalshift.html

[123] Maeno T. Origin and evolution of consciousness and emotion: has consciousness emerged from episodic memory?Emotions of animals and humans: comparative perspectives. Tokyo: Springer 2012; pp. 243-63.
[http://dx.doi.org/10.1007/978-4-431-54123-3_11]

[124] Kandel ER, Siegelbaum SA. Cellular mechanisms of implicit memory storage and biological basis of individuality.Principles of neural science. 5th ed. New York: McGraw Hill 2013; pp. 1461-85.

[125] Maier A, Adams GK, Aura C, Leopold D. Distinct superficial and deep laminar domains of activity in the visual cortex during rest and stimulation. 2010; 4(article 31): 1-11. Available from https://www.frontiersin.org/articles/10.3389/fnsys.2010.00031/full
[http://dx.doi.org/10.3389/fnsys.2010.00031]

[126] Fontolan L, Morillon B, Liegeois-Chauvel C, Giraud A-L. The contribution of frequency-specific

activity to hierarchical information processing in the human auditory cortex. Nat Commun 2014; 5(May): 4694.
[http://dx.doi.org/10.1038/ncomms5694] [PMID: 25178489]

[127] Anastassiou CA, Shai AS. Psyche, signals and systems.Micro-, meso- and macro-dynamics of the brain. Heidelberg: Springer 2016; pp. 99-156.
[http://dx.doi.org/10.1007/978-3-319-28802-4_9]

[128] Shai AS, Anastassiou CA, Larkum ME, Koch C. Physiology of layer 5 pyramidal neurons in mouse primary visual cortex: coincidence detection through bursting. PLOS Comput Biol 2015; 11(3): e1004090. Available from http://journals.plos.org/ploscompbiol/article?id=10.1371/journal.pcbi.1004090
[http://dx.doi.org/10.1371/journal.pcbi.1004090] [PMID: 25768881]

[129] Merrick C, Godwin CA, Geisler MW, Morsella E. The olfactory system as the gateway to the neural correlates of consciousness. Front Psychol 2014; 4(article 1011): 1-15. Available from https://www.ncbi.nlm.nih.gov/pmc/articles/PMC3887364/
[http://dx.doi.org/10.3389/fpsyg.2013.01011]

[130] Mori K, Manabe H. Unique characteristics of the olfactory system.The olfactory system: from odor molecules to motivational behaviors. Tokyo: Springer 2014; pp. 1-18.
[http://dx.doi.org/10.1007/978-4-431-54376-3_1]

[131] Touhara K. Odor and pheromone molecules, receptors, and behavioral responses.The olfactory system: from odor molecules to motivational behaviors. Tokyo: Springer 2014; pp. 19-38.
[http://dx.doi.org/10.1007/978-4-431-54376-3_2]

[132] Nagayama S, Igarashi KM, Manabe H, Mori K. Parallel tufted cell and mitral cell pathways from the olfactory bulb to olfactory cortex.The olfactory system: from odor molecules to motivational behaviors. Tokyo: Sieger 2014; pp. 133-60.
[http://dx.doi.org/10.1007/978-4-431-54376-3_7]

[133] Mori K. Piriform cortex and olfactory tubercle.The olfactory system: from odor molecules to motivational behaviors. Tokyo: Springer 2014; pp. 161-76.
[http://dx.doi.org/10.1007/978-4-431-54376-3_8]

[134] Singer W, Gray CM. Visual feature integration and the temporal correlation hypothesis. Annu Rev Neurosci 1995; 18: 555-86.
[http://dx.doi.org/10.1146/annurev.ne.18.030195.003011] [PMID: 7605074]

[135] Gray CM, König P, Engel AK, Singer W. Oscillatory responses in cat visual cortex exhibit inter-columnar synchronization which reflects global stimulus properties. Nature 1989; 338(6213): 334-7.
[http://dx.doi.org/10.1038/338334a0] [PMID: 2922061]

[136] Treisman A. Feature biniding, attention, and object perception.Essential sources in the scientific study of consciousness. Cambridge: The MIT Press 2003; pp. 63-83.

[137] Engel AK, Fries P, Roelfsema PR, König P, Brecht M, Singer W. Temporal binding, binocular rivaltry, and consciousness.Essential sources in the scientific study of consciousness. Cambridge: The MIT Press 2003; pp. 126-46.

[138] Munk MH, Roelfsema PR, König P, Engel AK, Singer W. Role of Reticular Activation in the Modulation of Intracortical Synchronization 1996; 272: 271-4.

[139] Wilson DA, Sullivan RM. Cortical processing of odor objects. Neuron 2011; 72(4): 506-19.
[http://dx.doi.org/10.1016/j.neuron.2011.10.027] [PMID: 22099455]

[140] Aboitiz F, Montiel JF. Olfaction, navigation, and the origin of isocortex. Front Neurosci 2015; 9(article 402): 1-12. Available from https://www.ncbi.nlm.nih.gov/pmc/articles/PMC4621927/
[http://dx.doi.org/10.3389/fnins.2015.00402]

[141] Weiss T, Secundo L, Sobel N. Human olfaction: a typical yet special mammalian olfactory system.The olfactory system: from odor molecules to motivational behaviors. Tokyo: Springer 2014; pp. 177-202.

[http://dx.doi.org/10.1007/978-4-431-54376-3_9]

[142] Zatorre RJ, Jones-Gotman M, Evans AC, Meyer E. Functional localization and lateralization of human olfactory cortex. Nature 1992; 360(6402): 339-40. [http://dx.doi.org/10.1038/360339a0] [PMID: 1448149]

[143] Qureshy A, Kawashima R, Imran MB, *et al.* Functional mapping of human brain in olfactory processing: a PET study. J Neurophysiol 2000; 84(3): 1656-66. [http://dx.doi.org/10.1152/jn.2000.84.3.1656] [PMID: 10980035]

[144] Kjelvik G, Evensmoen HR, Brezova V, Håberg AK. The human brain representation of odor identification. J Neurophysiol 2012; 108(2): 645-57. [http://dx.doi.org/10.1152/jn.01036.2010] [PMID: 22539820]

[145] Mizobuchi M, Ito N, Tanaka C, Sako K, Sumi Y, Sasaki T. Unidirectional olfactory hallucination associated with ipsilateral unruptured intracranial aneurysm. Epilepsia 1999; 40(4): 516-9. [http://dx.doi.org/10.1111/j.1528-1157.1999.tb00751.x] [PMID: 10219282]

[146] Li W, Lopez L, Osher J, Howard JD, Parrish TB, Gottfried JA. Right orbitofrontal cortex mediates conscious olfactory perception. Psychol Sci 2010; 21(10): 1454-63. [http://dx.doi.org/10.1177/0956797610382121] [PMID: 20817780]

[147] Wu KN, Tan BK, Howard JD, Conley DB, Gottfried JA. Olfactory input is critical for sustaining odor quality codes in human orbitofrontal cortex. Nat Neurosci 2012; 15(9): 1313-9. [http://dx.doi.org/10.1038/nn.3186] [PMID: 22885850]

[148] Gottfried JA, Winston JS, Dolan RJ. Dissociable codes of odor quality and odorant structure in human piriform cortex. Neuron 2006; 49(3): 467-79. [http://dx.doi.org/10.1016/j.neuron.2006.01.007] [PMID: 16446149]

[149] Mori K, Manabe H, Narikiyo K, Onisawa N. Olfactory consciousness and gamma oscillation couplings across the olfactory bulb, olfactory cortex, and orbitofrontal cortex. Front Psychol 2013. Available from https://www.frontiersin.org/articles/10.3389/fpsyg.2013.00743/full [http://dx.doi.org/10.3389/fpsyg.2013.00743]

[150] Howard JD, Plailly J, Grueschow M, Haynes J-D, Gottfried JA. Odor quality coding and categorization in human posterior piriform cortex. Nat Neurosci 2009; 12(7): 932-8. [http://dx.doi.org/10.1038/nn.2324] [PMID: 19483688]

[151] Yoshihara Y. Zebrafish olfactory system.The olfactory system: from odor molecules to motivational behaviors. Tokyo: Springer 2014; pp. 97-132. [http://dx.doi.org/10.1007/978-4-431-54376-3_5]

[152] Boyd AM, Sturgill JF, Poo C, Isaacson JS. Cortical feedback control of olfactory bulb circuits. Neuron 2012; 76(6): 1161-74. [http://dx.doi.org/10.1016/j.neuron.2012.10.020] [PMID: 23259951]

[153] Knaden M, Strutz A, Ahsan J, Sachse S, Hansson BS. Spatial representation of odorant valence in an insect brain. Cell Reports 2012; 1(4): 392-9. [http://dx.doi.org/10.1016/j.celrep.2012.03.002] [PMID: 22832228]

[154] Yeshurun Y, Sobel N. An odor is not worth a thousand words: from multidimensional odors to unidimensional odor objects. Annu Rev Psychol 2010; 61: 219-241, C1-C5. [http://dx.doi.org/10.1146/annurev.psych.60.110707.163639] [PMID: 19958179]

[155] Sela L, Sobel N. Human olfaction: a constant state of change-blindness. Exp Brain Res 2010; 205(1): 13-29. [http://dx.doi.org/10.1007/s00221-010-2348-6] [PMID: 20603708]

[156] Haddad R, Weiss T, Khan R, *et al.* Global features of neural activity in the olfactory system form a parallel code that predicts olfactory behavior and perception. J Neurosci 2010; 30(27): 9017-26. [http://dx.doi.org/10.1523/JNEUROSCI.0398-10.2010] [PMID: 20610736]

[157] Weiss T, Snitz K, Yablonka A, *et al.* Perceptual convergence of multi-component mixtures in olfaction implies an olfactory white. Proc Natl Acad Sci USA 2012; 109(49): 19959-64. Available from http://www.pnas.org/content/109/49/19959.long [http://dx.doi.org/10.1073/pnas.1208110109] [PMID: 23169632]

[158] Lapid H, Shushan S, Plotkin A, *et al.* Neural activity at the human olfactory epithelium reflects olfactory perception. Nat Neurosci 2011; 14(11): 1455-61. [http://dx.doi.org/10.1038/nn.2926] [PMID: 21946326]

[159] Oswald A-MM, Urban NN. Interactions between behaviorally relevant rhythms and synaptic plasticity alter coding in the piriform cortex. J Neurosci 2012; 32(18): 6092-104. [http://dx.doi.org/10.1523/JNEUROSCI.6285-11.2012] [PMID: 22553016]

[160] Khan RM, Luk C-H, Flinker A, *et al.* Predicting odor pleasantness from odorant structure: pleasantness as a reflection of the physical world. J Neurosci 2007; 27(37): 10015-23. [http://dx.doi.org/10.1523/JNEUROSCI.1158-07.2007] [PMID: 17855616]

[161] Tsuchiya N, Koch C. The relationship between consciousness and top-down attention.The neurology of consciousness: cognitive science and neuropathology. 2nd ed. San Diego: Academic Press 2016; pp. 71-91. [http://dx.doi.org/10.1016/B978-0-12-800948-2.00005-4]

[162] Grossberg S. Towards solving the hard problem of consciousness: The varieties of brain resonances and the conscious experiences that they support. Neural Netw 2017; 87: 38-95. [http://dx.doi.org/10.1016/j.neunet.2016.11.003] [PMID: 28088645]

[163] Mehta N, Mashour GA. General and specific consciousness: a first-order representationalist approach. Neural Netw 2013; 14(article 407): 1-9. Available from https://www.ncbi.nlm.nih.gov/pmc/articles/PMC3712269/ [http://dx.doi.org/10.3389/fpsyg.2013.00407]

[164] Melloni L, Molina C, Pena M, Torres D, Singer W, Rodriguez E. Synchronization of neural activity across cortical areas correlates with conscious perception. J Neurosci 2007; 27(11): 2858-65. [http://dx.doi.org/10.1523/JNEUROSCI.4623-06.2007] [PMID: 17360907]

[165] Doesburg SM, Green JJ, McDonald JJ, Ward LM. Rhythms of consciousness: binocular rivalry reveals large-scale oscillatory network dynamics mediating visual perception. PLoS One 2009; 4(7): e6142. Available from http://journals.plos.org/plosone/ [http://dx.doi.org/10.1371/journal.pone.0006142] [PMID: 19582165]

[166] Ribary U, Ioannides AA, Singh KD, *et al.* Magnetic field tomography of coherent thalamocortical 40-Hz oscillations in humans. Proc Natl Acad Sci USA 1991; 88(24): 11037-41. [http://dx.doi.org/10.1073/pnas.88.24.11037] [PMID: 1763020]

[167] Damasio AR. Synchronous activation in multiple cortical regions: a mechanism for recall. Semin Neurosci 1990; 2: 287-96.

[168] Alais D, Blake R, Lee S-H. Visual features that vary together over time group together over space. Nat Neurosci 1998; 1(2): 160-4. [http://dx.doi.org/10.1038/414] [PMID: 10195133]

[169] Desmedt JE, Tomberg C. Transient phase-locking of 40 Hz electrical oscillations in prefrontal and parietal human cortex reflects the process of conscious somatic perception. Neurosci Lett 1994; 168(1-2): 126-9. [http://dx.doi.org/10.1016/0304-3940(94)90432-4] [PMID: 8028764]

[170] Eckhorn R, Bauer R, Jordan W, *et al.* Coherent oscillations: a mechanism of feature linking in the visual cortex? Multiple electrode and correlation analyses in the cat. Biol Cybern 1988; 60(2): 121-30. [http://dx.doi.org/10.1007/BF00202899] [PMID: 3228555]

[171] Joliot M, Ribary U, Llinás R. Human oscillatory brain activity near 40 Hz coexists with cognitive temporal binding. Proc Natl Acad Sci USA 1994; 91(24): 11748-51.

[http://dx.doi.org/10.1073/pnas.91.24.11748] [PMID: 7972135]

[172] Engel AK, Fries P. Neuronal oscillations, coherence, and consciousness.The neurology of consciousness: cognitive science and neuropathology. 2nd ed. San Diego: Academic Press 2016; pp. 49-60.
[http://dx.doi.org/10.1016/B978-0-12-800948-2.00003-0]

[173] Havlík M. Missing piece of the puzzle in the science of consciousness: Resting state and endogenous correlates of consciousness. Conscious Cogn 2017; 49: 70-85.
[http://dx.doi.org/10.1016/j.concog.2017.01.006] [PMID: 28160667]

[174] Merker BH. Cortical gamma oscillations: details of their genesis preclude a role in cognition. Front Comput Neurosci 2016; 10(article 78): 1-5. Available from http://journal.frontiersin.org/article/10.3389/fncom.2016.00078
[http://dx.doi.org/10.3389/fncom.2016.00078]

[175] Prinz J. A neurofunctional theory of consciousness.Cognition and the brain: the philosophy and neuroscience movement. Cambridge: Cambridge University Press 2005.
[http://dx.doi.org/10.1017/CBO9780511610608.012]

[176] Davey CG, Pujol J, Harrison BJ. Mapping the self in the brain's default mode network. Neuroimage 2016; 132: 390-7.
[http://dx.doi.org/10.1016/j.neuroimage.2016.02.022] [PMID: 26892855]

[177] Hyafil A, Fontolan L, Kabdebon C, Gutkin B, Giraud A-L. Speech encoding by coupled cortical theta and gamma oscillations. eLife 2015; 4: e06213.
[http://dx.doi.org/10.7554/eLife.06213] [PMID: 26023831]

[178] Arnal LH, Giraud AL. Cortical oscillations and sensory predictions. Trends Cogn Sci (Regul Ed) 2012; 16(7): 390-8.
[http://dx.doi.org/10.1016/j.tics.2012.05.003] [PMID: 22682813]

[179] Poeppel D, Idsardi WJ, van Wassenhove V. Speech perception at the interface of neurobiology and linguistics. Philos Trans R Soc Lond B Biol Sci 2008; 363(1493): 1071-86. Available from http://rstb.royalsocietypublishing.org/content/363/1493/1071.long
[http://dx.doi.org/10.1098/rstb.2007.2160] [PMID: 17890189]

[180] Gagnepain P, Henson RN, Davis MH. Temporal predictive codes for spoken words in auditory cortex. Curr Biol 2012; 22(7): 615-21.
[http://dx.doi.org/10.1016/j.cub.2012.02.015] [PMID: 22425155]

[181] Schroeder CE, Lakatos P. Low-frequency neuronal oscillations as instruments of sensory selection. Trends Neurosci 2009; 32(1): 9-18. Available from https://elifesciences.org/articles/06213
[http://dx.doi.org/10.1016/j.tins.2008.09.012] [PMID: 19012975]

[182] Jerath R, Crawford MW. Layers of human brain activity: a functional model based on the default mode network and slow oscillations. Front Hum Neurosci 2015; 9(article 248): 1-16. Available from https://www.ncbi.nlm.nih.gov/pmc/articles/PMC4413559/
[http://dx.doi.org/10.3389/fnhum.2015.00248]

[183] Baars BJ. Multiple sources of conscious odor integration and propagation in olfactory cortex. Front Psychol 2013; 4(article 930): 1-4. Available from https://www.ncbi.nlm.nih.gov/pmc/articles/PMC3868888/
[http://dx.doi.org/10.3389/fpsyg.2013.00930]

[184] Faivre N, Mudrik L, Schwartz N, Koch C. Multisensory integration in complete unawareness: evidence from audiovisual congruency priming. Psychol Sci 2014; 25(11): 2006-16. Available from http://pss.sagepub.com/lookup/doi/10.1177/0956797614547916
[http://dx.doi.org/10.1177/0956797614547916] [PMID: 25269620]

[185] Mudrik L, Faivre N, Koch C. Information integration without awareness. Trends Cogn Sci (Regul Ed) 2014; 18(9): 488-96.

[http://dx.doi.org/10.1016/j.tics.2014.04.009] [PMID: 24933626]

[186] Zeki S. A theory of microconsciousness.The Blackwell companion to consciousness. Malden: Blackwell 2007; pp. 580-8.
[http://dx.doi.org/10.1002/9780470751466.ch46]

[187] Lau H, Rosenthal D. Empirical support for higher-order theories of conscious awareness. Trends Cogn Sci (Regul Ed) 2011; 15(8): 365-73.
[http://dx.doi.org/10.1016/j.tics.2011.05.009] [PMID: 21737339]

[188] Aru J, Bachmann T, Singer W, Melloni L. Distilling the neural correlates of consciousness. Neurosci Biobehav Rev 2012; 36(2): 737-46.
[http://dx.doi.org/10.1016/j.neubiorev.2011.12.003] [PMID: 22192881]

[189] Baars BJ, Franklin S, Ramsoy TZ. Global workspace dynamics: cortical "binding and propagation" enables conscious contents. Front Psychol 2013; 4(article 200): 1-22. Available from https://www.frontiersin.org/articles/10.3389/fpsyg.2013.00200/full

[190] Baars B. A Cognitive Theory of Consciousness. Cambridge UP 1998.

[191] Dehaene S, Kerszberg M, Changeux JP. A neuronal model of a global workspace in effortful cognitive tasks. Proc Natl Acad Sci USA 1998; 95(24): 14529-34.
[http://dx.doi.org/10.1073/pnas.95.24.14529] [PMID: 9826734]

[192] Koch C, Massimini M, Boly M, Tononi G. Neural correlates of consciousness: progress and problems. Nat Rev Neurosci 2016; 17(5): 307-21.
[http://dx.doi.org/10.1038/nrn.2016.22] [PMID: 27094080]

[193] Panagiotaropoulos TI, Deco G, Kapoor V, Logothetis NK. Neuronal discharges and gamma oscillations explicitly reflect visual consciousness in the lateral prefrontal cortex. Neuron 2012; 74(5): 924-35.
[http://dx.doi.org/10.1016/j.neuron.2012.04.013] [PMID: 22681695]

[194] Bachmann T. On the brain-imaging markers of neural correlates of consciousness. Front Psych 2015; 6(article 868): 1-4. Available from https://www.frontiersin.org/articles/10.3389/fpsyg.2015.00868/full
[http://dx.doi.org/10.3389/fpsyg.2015.00868]

[195] Crick F, Koch C. Are we aware of neural activity in primary visual cortex? Nature 1995; 375(6527): 121-3.
[http://dx.doi.org/10.1038/375121a0] [PMID: 7753166]

[196] Del Cul A, Baillet S, Dehaene S. Brain dynamics underlying the nonlinear threshold for access to consciousness. PLoS Biol 2007; 5(10): e260.
[http://dx.doi.org/10.1371/journal.pbio.0050260] [PMID: 17896866]

[197] Edelman GM, Gally JA. Reentry: a key mechanism for integration of brain function. Front Integr Nuerosci 2013; 7(63): 63.
[PMID: 23986665]

[198] Pins D, Ffytche D. The neural correlates of conscious vision. Cereb Cortex 2003; 13(5): 461-74.
[http://dx.doi.org/10.1093/cercor/13.5.461] [PMID: 12679293]

[199] Sheinberg DL, Logothetis NK. The role of temporal cortical areas in perceptual organization. Proc Natl Acad Sci USA 1997; 94(7): 3408-13.
[http://dx.doi.org/10.1073/pnas.94.7.3408] [PMID: 9096407]

[200] Haynes JD, Rees G. Predicting the orientation of invisible stimuli from activity in human primary visual cortex. Nat Neurosci 2005; 8(5): 686-91.
[http://dx.doi.org/10.1038/nn1445] [PMID: 15852013]

[201] Lee SH, Blake R, Heeger DJ. Hierarchy of cortical responses underlying binocular rivalry. Nat Neurosci 2007; 10(8): 1048-54.
[http://dx.doi.org/10.1038/nn1939] [PMID: 17632508]

[202] Dehaene S, Changeux JP, Naccache L, Sackur J, Sergent C. Conscious, preconscious, and subliminal processing: a testable taxonomy. Trends Cogn Sci (Regul Ed) 2006; 10(5): 204-11. [http://dx.doi.org/10.1016/j.tics.2006.03.007] [PMID: 16603406]

[203] Noy N, Bickel S, Zion-Golumbic E, *et al.* Ignition's glow: Ultra-fast spread of global cortical activity accompanying local "ignitions" in visual cortex during conscious visual perception. Conscious Cogn 2015; 35: 206-24. [http://dx.doi.org/10.1016/j.concog.2015.03.006] [PMID: 25824626]

[204] Block N. Consciousness, accessibility, and the mesh between psychology and neuroscience. Behav Brain Sci 2007; 30(5-6): 481-99. [http://dx.doi.org/10.1017/S0140525X07002786] [PMID: 18366828]

[205] Baars BJ. Global workspace theory of consciousness: toward a cognitive neuroscience of human experience. Prog Brain Res 2005; 150: 45-53. [http://dx.doi.org/10.1016/S0079-6123(05)50004-9] [PMID: 16186014]

[206] Kahneman D, Tversky A. Prospect Theory: an analysis of decision under risk. Econometrica 1979; 47(3): 263-91. [http://dx.doi.org/10.2307/1914185]

[207] Nicolelis M. Beyond boundaries: the new neuroscience of connecting brains with machines - and how it will change our lives. New York: St Martin's Griffin 2011.

[208] Vanhatalo S, Palva JM, Holmes MD, Miller JW, Voipio J, Kaila K. Infraslow oscillations modulate excitability and interictal epileptic activity in the human cortex during sleep. Proc Natl Acad Sci USA 2004; 101(14): 5053-7. [http://dx.doi.org/10.1073/pnas.0305375101] [PMID: 15044698]

[209] Monto S, Palva S, Voipio J, Palva JM. Very slow EEG fluctuations predict the dynamics of stimulus detection and oscillation amplitudes in humans. J Neurosci 2008; 28(33): 8268-72. [http://dx.doi.org/10.1523/JNEUROSCI.1910-08.2008] [PMID: 18701689]

[210] Jerath R, Crawford MW, Barnes VA. A unified 3D default space consciousness model combining neurological and physiological processes that underlie conscious experience. Front Psychol 2015; 6(article 1204): 1-25. Available from http://www.pubmedcentral.nih.gov/articlerender.fcgi?artid=4550793&tool=pmcentrez&rendertype=abstract

[211] Fox MD, Snyder AZ, Vincent JL, Corbetta M, Van Essen DC, Raichle ME. The human brain is intrinsically organized into dynamic, anticorrelated functional networks. Proc Natl Acad Sci USA 2005; 102(27): 9673-8. Available from http://www.pubmedcentral.nih.gov/articlerender.fcgi?artid=1157105&tool=pmcentrez&rendertype=abstract [http://dx.doi.org/10.1073/pnas.0504136102] [PMID: 15976020]

[212] Newman J, Baars BJ. A neural attentional model of access to consciousness: a global workspace perspective. Concepts Neurosci 1993; 4: 255-90.

[213] Dehaene S, Kerszberg M, Changeux J-P. A neuronal model of a global workspace in effortful cognitive tasks. Proc Natl Acad Sci USA 1998; 95(24): 14529-34. [http://dx.doi.org/10.1073/pnas.95.24.14529] [PMID: 9826734]

[214] Baars BJ. The conscious access hypothesis: origins and recent evidence. Trends Cogn Sci (Regul Ed) 2002; 6(1): 47-52. [http://dx.doi.org/10.1016/S1364-6613(00)01819-2] [PMID: 11849615]

[215] Shanahan M. A spiking neuron model of cortical broadcast and competition. Conscious Cogn 2008; 17(1): 288-303. [http://dx.doi.org/10.1016/j.concog.2006.12.005] [PMID: 17317220]

[216] Dehaene S, Changeux JP. Experimental and theoretical approaches to conscious processing. Neuron 2011; 70(2): 200-27. Available from https://www.cell.com/neuron/fulltext/S0896-6273(11)00258-3 [http://dx.doi.org/10.1016/j.neuron.2011.03.018] [PMID: 21521609]

[217] Dehaene S, Charles L, King JR, Marti S. Toward a computational theory of conscious processing. Curr Opin Neurobiol 2014; 25: 76-84.
[http://dx.doi.org/10.1016/j.conb.2013.12.005] [PMID: 24709604]

[218] Edelman GM, Gally JA, Baars BJ. Biology of consciousness. Front Psychol 2011; 2(article 4): 1-7. Available from https://www.frontiersin.org/articles/10.3389/fpsyg.2011.00004/full
[http://dx.doi.org/10.3389/fpsyg.2011.00004]

[219] Barttfeld P, Uhrig L, Sitt JD, Sigman M, Jarraya B, Dehaene S. Signature of consciousness in the dynamics of resting-state brain activity. Proc Natl Acad Sci USA 2015; 112(3): 887-92. Available from www.pnas.org/cgi/doi/10.1073/pnas.1418031112
[http://dx.doi.org/10.1073/pnas.1418031112] [PMID: 25561541]

[220] Chennu S, Finoia P, Kamau E, *et al.* Spectral signatures of reorganised brain networks in disorders of consciousness. PLOS Comput Biol 2014; 10(10): e1003887. Available from http://journals.plos.org/ploscompbiol/article?id=10.1371/journal.pcbi.1003887
[http://dx.doi.org/10.1371/journal.pcbi.1003887] [PMID: 25329398]

[221] Amaral D. The functional organization of perception and movement.Principles of neural science. 5th ed. New York: McGraw Hill 2013; pp. 356-69.

[222] Carhart-Harris RL, Friston KJ. The default-mode, ego-functions and free-energy: a neurobiological account of Freudian ideas. Brain 2010; 133(Pt 4): 1265-83.
[http://dx.doi.org/10.1093/brain/awq010] [PMID: 20194141]

[223] Leske S, Ruhnau P, Frey J, *et al.* Prestimulus network integration of auditory cortex predisposes near-threshold perception independently of local excitability. Cereb Cortex 2015; 25(12): 4898-907.
[http://dx.doi.org/10.1093/cercor/bhv212] [PMID: 26408799]

[224] Klimesch W. An algorithm for the EEG frequency architecture of consciousness and brain body coupling. Front Hum Neurosci 2013; 7(article 766): 1-4. Available from https://www.ncbi.nlm.nih.gov/pmc/articles/PMC3824085/
[http://dx.doi.org/10.3389/fnhum.2013.00766]

[225] Freeman WJ, Burke BC, Holmes MD. Aperiodic phase re-setting in scalp EEG of beta-gamma oscillations by state transitions at alpha-theta rates. Hum Brain Mapp 2003; 19(4): 248-72.
[http://dx.doi.org/10.1002/hbm.10120] [PMID: 12874778]

[226] Steriade M. Grouping of brain rhythms in corticothalamic systems. Neuroscience 2006; 137(4): 1087-106.
[http://dx.doi.org/10.1016/j.neuroscience.2005.10.029] [PMID: 16343791]

[227] Uttal WR. Mind and brain: a critical appraisal of cognitive neuroscience. Cambridge: MIT Press 2011; pp. 1-47.
[http://dx.doi.org/10.7551/mitpress/9780262015967.001.0001]

[228] Weisz N, Wühle A, Monittola G, *et al.* Prestimulus oscillatory power and connectivity patterns predispose conscious somatosensory perception. Proc Natl Acad Sci USA 2014; 111(4): E417-25.
[http://dx.doi.org/10.1073/pnas.1317267111] [PMID: 24474792]

[229] Gonzalez-Castillo J, Saad ZS, Handwerker DA, Inati SJ, Brenowitz N, Bandettini PA. Whole-brain, time-locked activation with simple tasks revealed using massive averaging and model-free analysis. Proc Natl Acad Sci USA 2012; 109(14): 5487-92. Available from http://www.pnas.org/content/109/14/5487.long
[http://dx.doi.org/10.1073/pnas.1121049109] [PMID: 22431587]

[230] Raichle ME. Two views of brain function. Trends Cogn Sci (Regul Ed) 2010; 14(4): 180-90.
[http://dx.doi.org/10.1016/j.tics.2010.01.008] [PMID: 20206576]

[231] Demertzi A, Whitfield-Gabrieli S. Intrinsic brain activity and consciousness.The neurology of consciousness: cognitive science and neuropathology. 2nd ed. San Diego: Academic Press 2016; pp. 95-105.

[http://dx.doi.org/10.1016/B978-0-12-800948-2.00006-6]

[232] van den Heuvel MP, Mandl RC, Kahn RS, Hulshoff Pol HE. Functionally linked resting-state networks reflect the underlying structural connectivity architecture of the human brain. Hum Brain Mapp 2009; 30(10): 3127-41.
[http://dx.doi.org/10.1002/hbm.20737] [PMID: 19235882]

[233] Crone JS, Schurz M, Holler Y, *et al.* Impaired consciousness is linked to changes in effective connectivity of the posterior cingulate cortex. Neuroimage 2015; 110(April 15): 101-9.

[234] Utevsky AV, Smith DV, Huettel SA. Precuneus is a functional core of the default-mode network. J Neurosci 2014; 34(3): 932-40.
[http://dx.doi.org/10.1523/JNEUROSCI.4227-13.2014] [PMID: 24431451]

[235] Horovitz SG, Braun AR, Carr WS, *et al.* Decoupling of the brain's default mode network during deep sleep. Proc Natl Acad Sci USA 2009; 106(27): 11376-81.
[http://dx.doi.org/10.1073/pnas.0901435106] [PMID: 19549821]

[236] Graziano MS, Kastner S. Human consciousness and its relationship to social neuroscience: A novel hypothesis. Cogn Neurosci 2011; 2(2): 98-113.
[http://dx.doi.org/10.1080/17588928.2011.565121] [PMID: 22121395]

[237] Hudetz AG. General anesthesia and human brain connectivity. Brain Connect 2012; 2(6): 291-302.
[http://dx.doi.org/10.1089/brain.2012.0107] [PMID: 23153273]

[238] Leighton TD. Cultivating the empty field: the silent illumination of Zen Master Hongzhi North Clarendon. Tuttle Publishing 2000.

[239] Britton WB, Lindahl JR, Cahn BR, Davis JH, Goldman RE. Awakening is not a metaphor: the effects of Buddhist meditation practices on basic wakefulness. Ann N Y Acad Sci 2014; 1307: 64-81.
[http://dx.doi.org/10.1111/nyas.12279] [PMID: 24372471]

[240] Uchiyama K. Opening the hand of thought Somerville. Wisdom Publications 2004.

[241] Uhrig L, Dehaene S, Jarraya B. Cerebral mechanisms of general anesthesia. Ann Fr Anesth Reanim 2014; 33(2): 72-82.
[http://dx.doi.org/10.1016/j.annfar.2013.11.005] [PMID: 24368069]

[242] Mashour GA. Consciousness and anesthesia.The neurology of consciousness: cognitive science and neuropathology. 2nd ed. San Diego: Academic Press 2016; pp. 139-52.
[http://dx.doi.org/10.1016/B978-0-12-800948-2.00009-1]

[243] Prinz J. The conscious brain: how attention engenders experience. New York: Oxford University Press 2012.
[http://dx.doi.org/10.1093/acprof:oso/9780195314595.001.0001]

[244] Safavi S, Kapoor V, Logothetis NK, Panagiotaropoulos TI. Is the frontal lobe involved in conscious perception? Front Psychol 2014; 5: 1063.
[http://dx.doi.org/10.3389/fpsyg.2014.01063] [PMID: 25285089]

[245] Hudetz AG. General anesthesia and human brain connectivity. Brain Connect 2012; 2(6): 291-302.
[http://dx.doi.org/10.1089/brain.2012.0107] [PMID: 23153273]

[246] Boveroux P, Vanhaudenhuyse A, Bruno MA, *et al.* Breakdown of within- and between-network resting state functional magnetic resonance imaging connectivity during propofol-induced loss of consciousness. Anesthesiology 2010; 113(5): 1038-53.
[http://dx.doi.org/10.1097/ALN.0b013e3181f697f5] [PMID: 20885292]

[247] Fischer DB, Boes AD, Demertzi A, *et al.* A human brain network derived from coma-causing brainstem lesions. Neurology 2016; 87(23): 2427-34.
[http://dx.doi.org/10.1212/WNL.0000000000003404] [PMID: 27815400]

[248] Hashmi JA, Loggia ML, Khan S, *et al.* Dexmedetomidine disrupts the local and global efficiency of large-scale brain networks. Anesthesiology 2017; 126(3): 419-30.

[http://dx.doi.org/10.1097/ALN.0000000000001509] [PMID: 28092321]

[249] Ku S-W, Lee U, Noh G-J, Jun IG, Mashour GA. Preferential inhibition of frontal-to-parietal feedback connectivity is a neurophysiologic correlate of general anesthesia in surgical patients. PLoS One 2011; 6(10): e25155. Available from http://journals.plos.org/plosone/article?id=10.1371/journal.pone.0025155
[http://dx.doi.org/10.1371/journal.pone.0025155] [PMID: 21998638]

[250] Jordan D, Ilg R, Riedl V, *et al.* Simultaneous electroencephalographic and functional magnetic resonance imaging indicate impaired cortical top-down processing in association with anesthetic-induced unconsciousness. Anesthesiology 2013; 119(5): 1031-42.
[http://dx.doi.org/10.1097/ALN.0b013e3182a7ca92] [PMID: 23969561]

[251] Lee U, Ku S, Noh G, Baek S, Choi B, Mashour GA. Disruption of frontal-parietal communication by ketamine, propofol, and sevoflurane. Anesthesiology 2013; 118(6): 1264-75.
[http://dx.doi.org/10.1097/ALN.0b013e31829103f5] [PMID: 23695090]

[252] Imas OA, Ropella KM, Ward BD, Wood JD, Hudetz AG. Volatile anesthetics disrupt frontal-posterior recurrent information transfer at gamma frequencies in rat. Neurosci Lett 2005; 387(3): 145-50.
[http://dx.doi.org/10.1016/j.neulet.2005.06.018] [PMID: 16019145]

[253] Liu X, Yanagawa T, Leopold DA, *et al.* Arousal transitions in sleep, wakefulness, and anesthesia are characterized by an orderly sequence of cortical events. Neuroimage 2015; 116: 222-31.
[http://dx.doi.org/10.1016/j.neuroimage.2015.04.003] [PMID: 25865143]

[254] Lee U, Ku S, Noh G, Baek S, Choi B, Mashour GA. Disruption of frontal-parietal communication by ketamine, propofol, and sevoflurane. Anesthesiology 2013; 118(6): 1264-75.
[http://dx.doi.org/10.1097/ALN.0b013e31829103f5] [PMID: 23695090]

[255] Meyer K. The role of dendritic signaling in the anesthetic suppression of consciousness. Anesthesiology 2015; 122(6): 1415-31.
[http://dx.doi.org/10.1097/ALN.0000000000000673] [PMID: 25901843]

[256] Craddock TJA, Kurian P, Preto J, *et al.* Anesthetic alterations of collective terahertz oscillations in tubulin correlate with clinical potency: implications for anesthetic action and post-operative cognitive dysfunction. Sci Rep 2017; 7(1): 9877. Available from https://www.nature.com/articles/s41598-01--09992-7
[http://dx.doi.org/10.1038/s41598-017-09992-7] [PMID: 28852014]

[257] Fang M, Tao Y-X, He F, *et al.* Synaptic PDZ domain-mediated protein interactions are disrupted by inhalational anesthetics. J Biol Chem 2003; 278(38): 36669-75.
[http://dx.doi.org/10.1074/jbc.M303520200] [PMID: 12853458]

[258] Bademosi AT, Steeves J, Karunanithi S, *et al.* Trapping of syntaxin 1a in presynaptic nanoclusters by a clinically relevant general anesthetic. Cell Reports 2018; 22(2): 427-40.
[http://dx.doi.org/10.1016/j.celrep.2017.12.054] [PMID: 29320738]

[259] Morsella E, Krieger SC, Bargh JA. Minimal neuroanatomy for a conscious brain: homing in on the networks constituting consciousness. Neural Netw 2010; 23(1): 14-5.
[http://dx.doi.org/10.1016/j.neunet.2009.08.004] [PMID: 19716260]

[260] Merker B. Consciousness without a cerebral cortex: a challenge for neuroscience and medicine. Behav Brain Sci 2007; 30(1): 63-81.
[http://dx.doi.org/10.1017/S0140525X07000891] [PMID: 17475053]

[261] Merker BH. The sentinel hypothesis: a role for the mammalian superior colliculus 1980.

[262] Fabbro F, Aglioti SM, Bergamasco M, *et al.* Evolutionary aspects of self- and world consciousness in vertebrates. Front Hum Neurosci 2015; 9(article 157): 1-16.

[263] Goddard CA, Sridharan D, Huguenard JR, Knudsen EI. Gamma oscillations are generated locally in an attention-related midbrain network. Neuron 2012; 73(3): 567-80.
[http://dx.doi.org/10.1016/j.neuron.2011.11.028] [PMID: 22325207]

[264] Bryant AS, Goddard CA, Huguenard JR, Knudsen EI. Cholinergic control of gamma power in the midbrain spatial attention network. J Neurosci 2015; 35(2): 761-75. [http://dx.doi.org/10.1523/JNEUROSCI.4001-14.2015] [PMID: 25589769]

[265] Goddard CA, Huguenard J, Knudsen E. Parallel midbrain microcircuits perform independent temporal transformations. J Neurosci 2014; 34(24): 8130-8. Available from http://www.ncbi.nlm.nih.gov/pubmed/24920618 [http://dx.doi.org/10.1523/JNEUROSCI.4399-13.2014] [PMID: 24920618]

[266] Stitt I, Galindo-Leon E, Pieper F, *et al.* Intrinsic coupling modes reveal the functional architecture of cortico-tectal networks. Sci Adv 2015; 1(7): e1500229. Available from http://www.pubmedcentral.nih.gov/articlerender.fcgi?artid=4643805&tool=pmcentrez&rendertype=abstract [http://dx.doi.org/10.1126/sciadv.1500229] [PMID: 26601226]

[267] Finkel LH, Edelman GM. Integration of distributed cortical systems by reentry: a computer simulation of interactive functionally segregated visual areas. J Neurosci 1989; 9(9): 3188-208. [http://dx.doi.org/10.1523/JNEUROSCI.09-09-03188.1989] [PMID: 2795160]

[268] Lamme VA, Roelfsema PR. The distinct modes of vision offered by feedforward and recurrent processing. Trends Neurosci 2000; 23(11): 571-9. [http://dx.doi.org/10.1016/S0166-2236(00)01657-X] [PMID: 11074267]

[269] Merker B. The efference cascade, consciousness, and its self: naturalizing the first person pivot of action control. Front Psychol 2013; 4(article 501): 1-20. Available from https://www.ncbi.nlm.nih.gov/pmc/articles/PMC3738861/ [http://dx.doi.org/10.3389/fpsyg.2013.00501]

[270] Halliwell JJ. Somewhere in the universe- where is the information stored when decohere? Phys Rev D Part Fields 1999; 102: 707-17.

[271] Craddock TJ, Priel A, Tuszynski JA. Keeping time: could quantum beating in microtubules be the basis for the neural synchrony related to consciousness? J Integr Neurosci 2014; 13(2): 293-311. [http://dx.doi.org/10.1142/S0219635214400019] [PMID: 25012713]

[272] Hales CG. The origins of the brain's endogenous electromagnetic field and its relationship to provision of consciousness. J Integr Neurosci 2014; 13(2): 313-61. [http://dx.doi.org/10.1142/S0219635214400056] [PMID: 25012714]

[273] Kafatos MC, Chevalier G, Chopra D, Hubacher J, Kak S, Theise ND. Biofield science: current physics perspective. Glob Adv Health Med 2015; 4 (Suppl.): 25-34. [http://dx.doi.org/10.7453/gahmj.2015.011.suppl] [PMID: 26665039]

[274] Sorli A, Dobnikar U, Fiscaletti D, Koroli V. Advanced relativity : multidimensionality of consciousness and mind, origin of life, psi phenomena. Neuroquantology 2017; 15(2): 109-17. [http://dx.doi.org/10.14704/nq.2017.15.2.1020]

[275] Cacha LA, Poznanski RR. Genomic instantiation of consciousness in neurons through a biophoton field theory. J Integr Neurosci 2014; 13(2): 253-92. [http://dx.doi.org/10.1142/S0219635214400081] [PMID: 25012712]

[276] Kurakin A. The self-organizing fractal theory as a universal discovery method: the phenomenon of life. Theor Biol Med Model 2011; 8(4): 4. Available from http://www.tbiomed.com/content/8/1/4 [http://dx.doi.org/10.1186/1742-4682-8-4] [PMID: 21447162]

[277] Bieberich E. Introduction to the fractality principle of consciousness and the sentyon postulate. Cognit Comput 2012; 4(1): 13-28. [http://dx.doi.org/10.1007/s12559-011-9104-5] [PMID: 23950765]

[278] Wheeler JA. Information, physics, quantum: the search for links. Complexity, entropy and the physics of information. Redwood City: Addison-Wesley Publishing Company 1989; pp. 3-28.

[279] Reimann MW, Nolte M, Scolamiero M, *et al.* Cliques of neurons bound into cavities provide a missing

link between structure and function. Front Comput Neurosci 2017; 11(article 48): 1-16. Available from https://www.frontiersin.org/articles/10.3389/fncom.2017.00048/full [http://dx.doi.org/10.3389/fncom.2017.00048]

[280] Gray JA. Consciousness: creeping up on the Hard Problem. New York, NY: Oxford University Press 2004.

[281] Taneichi T. Fermion field in the vicinity of a brane. Nat Sci 2013; 5(10): 1127-31.

[282] Taneichi T. Physical consciousness outside the brain: parasite fermion model for substance of consciousness. Neuroquantology 2015; 13(2): 134-40. [http://dx.doi.org/10.14704/nq.2015.13.2.838]

[283] Mender D. Isoqualitative gauge curvature at multiple scales: a response to the agnosticism of quantum cognitivism ad quantum interaction. Neuroquantology 2016; 14(4): 756-63. [http://dx.doi.org/10.14704/nq.2016.14.4.988]

[284] Morsella E, Godwin CA, Jantz TK, Krieger SC, Gazzaley A. Homing in on consciousness in the nervous system: An action-based synthesis. Behav Brain Sci 2016; 39: e168. Available from https://www.cambridge.org/core/journals/behavioral-and-brain-sciences/article/homing-in-on-consciousness-in-the-nervous-system-an-actionbased-synthesis/2483CA8F40A087A0A7AAABD40E0D89B2 [http://dx.doi.org/10.1017/S0140525X15000643] [PMID: 26096599]

[285] Benda J. The treason of the intellectuals. New York: Routledge 2017. La Trahison des Clercs [William Morrow & Company 1928] [http://dx.doi.org/10.4324/9781351298605]

[286] Thought experiment [cited 1st Sep 2017] Available from http://en.wikipedia.org/wiki/Thought_experiment

[287] Koch C. Consciousness: confessions of a romantic reductionist. Cambridge, London: The MIT Press 2012.

[288] Moran D. Introduction to phenomenology. London: Routledge 2000.

[289] Posner MI, Klein RM. On the functions of consciousness.Attention and performance IV. New York: Academic Press 1973.

[290] Variable [cited: 1st Sep 2017] Available from http://en.wikipedia.org/wiki/variable

[291] Collins Cobuild English Language Dictionary. London: Collins 1987.

[292] Conlon J. Why string theory?. Boca Raton: CRC Press 2016.

[293] Lookup table [cited: 1st Sep 2017]. Available from https://en.wikipedia.org/wiki/Lookup_table

[294] Axiom [cited 1st Jan 2016]. Available from http://en.wikipedia.org/wiki/Axiom

[295] Penrose SR. The quantum nature of consciousness 2013. [cited 1st Sep 2017]. Available from http://www.youtube.com/watch?v=3WXTX0IUaOg

[296] Paun G, Rozenberg G, Salomaa A. DNA computing: new computing paradigms. 2nd ed., Berlin: Springer 2005.

[297] de la Fuente IM. Metabolic dissipative structures.Systems biology of metabolic and signaling network. Heidelberg: Springer 2014; pp. 213-42. [http://dx.doi.org/10.1007/978-3-642-38505-6_8]

[298] Arkin A, Ross J. Computational functions in biochemical reaction networks. Biophys J 1994; 67(2): 560-78. [http://dx.doi.org/10.1016/S0006-3495(94)80516-8] [PMID: 7948674]

[299] Aon MA. Complex systems biology of networks: the riddle and the challenge.Systems biology of metabolic and signaling network. Heidelberg: Springer 2014; pp. 19-38.

[http://dx.doi.org/10.1007/978-3-642-38505-6_2]

[300] Aon MA, Lloyd D, Saks V. From physiology, genomes, systems, and self-organization to systems biology: the historica roots of a twenty-first century approach to complexity.Systems biology of metabolic and signaling network. Heidelberg: Springer 2014; pp. 3-17. [http://dx.doi.org/10.1007/978-3-642-38505-6_1]

[301] Scott AG. Nonlinear biology.The nonlinear universe: chaos, emergence, life. Berlin: Springer 2007; pp. 181-276.

[302] Darwin C. The origin of species. 2nd ed., New York: Bantam Dell 2008.

[303] Tolman EC. Cognitive maps in rats and men. Psychol Rev 1948; 55(4): 189-208. [http://dx.doi.org/10.1037/h0061626] [PMID: 18870876]

[304] Deroy O, Faivre N, Lunghi C, Spence C, Aller M, Noppeney U. The complex interplay between multisensory integration and perceptual awareness. Multisens Res 2016; 29(6-7): 585-606. Available from http://booksandjournals.brillonline.com/content/journals/10.1163/22134808-00002529 [http://dx.doi.org/10.1163/22134808-00002529] [PMID: 27795942]

[305] Merchant H, Zarco W, Pérez O, Prado L, Bartolo R. Measuring time with different neural chronometers during a synchronization-continuation task. Proc Natl Acad Sci USA 2011; 108(49): 19784-9. Available from http://www.pnas.org/content/108/49/19784.long [http://dx.doi.org/10.1073/pnas.1112933108] [PMID: 22106292]

[306] Crick F, Koch C. Towards a neurobiological theory of consciousness. Semin Neurosci 1990; 2: 263-75.

[307] Gould SJ. The structure of evolutionary theory. 4th., Cambrigde and London: The Belknar Press of Harvard University Press 2002.

[308] Merchant H, Bartolo R, Perez O, *et al.* Neurophysiology of timing in the hundreds of milliseconds: multiple layers of neuronal clocks in the medial premotor areas.Neurobiology of interval timing. New York: Springer 2014; pp. 143-54. [http://dx.doi.org/10.1007/978-1-4939-1782-2_8]

[309] Gray JA. The contents of consciousness: a neuropsychological conjecture. Behav Brain Sci 1995; 18: 659-76. [http://dx.doi.org/10.1017/S0140525X00040395]

[310] Roopun AK, Middleton SJ, Cunningham MO, *et al.* A beta2-frequency (20-30 Hz) oscillation in nonsynaptic networks of somatosensory cortex. Proc Natl Acad Sci USA 2006; 103(42): 15646-50. Available from http://www.pnas.org/content/103/42/15646.long [http://dx.doi.org/10.1073/pnas.0607443103] [PMID: 17030821]

[311] Wegner DM. The illusion of conscious will. Cambridge: MIT Press 2002.

[312] Koch C, Crick F. The zombie within. Nature 2001; 411(6840): 893. [http://dx.doi.org/10.1038/35082161] [PMID: 11418835]

[313] Neisser U. Cognitive psychology. New York: Appleton-Century-Crofts 1967.

[314] Panagiotaropoulos TI, Kapoor V, Logothetis NK. Desynchronization and rebound of beta oscillations during conscious and unconscious local neuronal processing in the macaque lateral prefrontal cortex. Front Psychol 2013; 4: 603. [http://dx.doi.org/10.3389/fpsyg.2013.00603] [PMID: 24062706]

[315] Bisenius S, Trapp S, Neumann J, Schroeter ML. Identifying neural correlates of visual consciousness with ALE meta-analyses. Neuroimage 2015; 122: 177-87. [http://dx.doi.org/10.1016/j.neuroimage.2015.07.070] [PMID: 26241685]

[316] Bogen JE. On the neurophysiology of consciousness: I. An overview. Conscious Cogn 1995; 4(1): 52-62. [http://dx.doi.org/10.1006/ccog.1995.1003] [PMID: 7497102]

[317] Llinás R, Ribary U, Joliot M, Wang X-J. Content and context in temporal thalamocortical binding.Temporal Coding in the Brain. Berlin: Springer 1994. [http://dx.doi.org/10.1007/978-3-642-85148-3_14]

[318] Schrouff J, Perlbarg V, Boly M, *et al.* Brain functional integration decreases during propofol-induced loss of consciousness. Neuroimage 2011; 57(1): 198-205. [http://dx.doi.org/10.1016/j.neuroimage.2011.04.020] [PMID: 21524704]

[319] Block N. Perceptual consciousness overflows cognitive access. Trends Cogn Sci (Regul Ed) 2011; 15(12): 567-75. [http://dx.doi.org/10.1016/j.tics.2011.11.001] [PMID: 22078929]

[320] Mangan B. Taking phenomenology seriously: the “Fringe” and its implications for cognitive research. Conscious Cogn 1993; 2(2): 89-108. [http://dx.doi.org/10.1006/ccog.1993.1008]

[321] Mangan B. The conscious “Fringe”: bringing William James up to date.Essential sources in the scientific study of consciousness. Cambridge: The MIT Press 2003; pp. 741-59.

[322] Norman E. Metacognition and mindfulness: the role of Fringe consciousness. Mindfulness (N Y) 2017; 8(1): 95-100. [http://dx.doi.org/10.1007/s12671-016-0494-z] [PMID: 28163796]

[323] Siegelbaum SA, Kandel ER, Yuste R. Synaptic integration in the central nervous system.Principles of neural science. 5th ed. New York: McGraw Hill 2013; pp. 210-35.

[324] Jessell TM, Sanes JR. Patterning the Nervous system.Principles of neural science. 5th ed. New York: McGraw Hill 2013; pp. 1165-86.

[325] Nagel T. What is like to be a bat? Philos Rev 1974; 83(4): 435-50. [http://dx.doi.org/10.2307/2183914]

[326] Sprague JM. James M Sprague.The history of neuroscience in autobiography. Washigton, DC: Society for Neuroscience 1996; Vol. 1: pp. 498-527.

[327] Weddell RA. Subcortical modulation of spatial attention including evidence that the Sprague effect extends to man. Brain Cogn 2004; 55(3): 497-506. [http://dx.doi.org/10.1016/j.bandc.2004.02.075] [PMID: 15223196]

[328] Chan JL, Koval MJ, Womelsdorf T, Lomber SG, Everling S. Dorsolateral prefrontal cortex deactivation in monkeys reduces preparatory beta and gamma power in the superior colliculus. Cereb Cortex 2015; 25(12): 4704-14. [http://dx.doi.org/10.1093/cercor/bhu154] [PMID: 25037923]

[329] Penfield W. Mystery of the mind: a critical study of consciousness and the human brain. 1st ed., Princeton: Princeton University Press 1975.

[330] George EP. [cited: 1st Jan 2017]. Available from https://en.m.wikiquote.org/wiki/George_E._P._Box

[331] Smith JB, Liang Z, Watson GD, *et al.* Interhemispheric resting-state functional connectivity of the claustrum in the awake and anesthetized states. Brain Structure and Function 2017; 222(5): 2041-58. [http://dx.doi.org/10.1007/s00429-016-1323-9]

SUBJECT INDEX

A

B

C

D

E

H

I

N

O

P

Q

T

U

V

W

www.ingramcontent.com/pod-product-compliance
Lightning Source LLC
LaVergne TN
LVHW070949210926
848482LV00002B/156